"十四五"职业教育国家规划教材

"十三五"职业教育国家规划教材

高等职业教育计算机系列教材

Android Studio 移动应用开发基础
（第2版）

吴绍根　罗　佳　主　编

电子工业出版社

Publishing House of Electronics Industry

北京·BEIJING

内 容 简 介

本书使用通俗易懂的语言讲解 Android 的基本理论知识,并且结合大量简单易懂的案例引导、帮助读者理解和掌握 Android 的重要知识点及应用技巧。本书的主要内容包括 Android 概述、Android 界面开发、Android 数据存储、Android 多媒体开发及网络开发等,还介绍了 Android 与 HTML5 的混合开发技术。本书针对各章节涉及的知识点,安排了多个案例,用于引导读者学习,由易到难,循序渐进。编者通过逐步操作案例,介绍知识点的应用情况,同时,针对每个案例设计对应的练习题,让读者能够对所学知识点进行应用、实践。

本书内容翔实,案例经典,实践性强,既可以作为高职、高专移动互联网应用技术专业课程的教材和教学参考书,又可以供从事 Android 移动应用开发工作的用户学习和参考。无论是拥有丰富 Java 开发经验的程序员,还是只有 Java 基础的初学者,本书都是十分有价值的学习资料。

未经许可,不得以任何方式复制或抄袭本书之部分或全部内容。
版权所有,侵权必究。

图书在版编目(CIP)数据

Android Studio 移动应用开发基础 / 吴绍根,罗佳主编. —2 版. —北京:电子工业出版社,2023.2

ISBN 978-7-121-44852-2

Ⅰ. ①A… Ⅱ. ①吴… ②罗… Ⅲ. ①移动终端－应用程序－程序设计－高等学校－教材 Ⅳ. ①TN929.53

中国国家版本馆 CIP 数据核字(2023)第 005424 号

责任编辑:徐建军　　　　　　　　特约编辑:田学清
印　　刷:河北鑫兆源印刷有限公司
装　　订:河北鑫兆源印刷有限公司
出版发行:电子工业出版社
　　　　　北京市海淀区万寿路 173 信箱　　　邮编:100036
开　　本:787×1 092　　1/16　　印张:18.25　　字数:491 千字
版　　次:2019 年 8 月第 1 版
　　　　　2023 年 2 月第 2 版
印　　次:2025 年 8 月第 9 次印刷
定　　价:55.00 元

凡所购买电子工业出版社图书有缺损问题,请向购买书店调换。若书店售缺,请与本社发行部联系,联系及邮购电话:(010)88254888,88258888。
质量投诉请发邮件至 zlts@phei.com.cn,盗版侵权举报请发邮件至 dbqq@phei.com.cn。
本书咨询联系方式:(010)88254570,xujj@phei.com.cn。

前　言

　　本书是一本介绍 Android 应用开发基础的实用教材，全面介绍 Android 应用开发的基础知识。在保持第 1 版通俗、易懂、易学特点的同时，第 2 版对 Android 的重要知识点和案例重新进行梳理，增加 Android 和 HTML5 混合开发的相关知识和案例。本书采用最新的 Android Studio 开发工具，语言通俗、易懂，操作步骤详细，编程思路清晰。本书知识点的组织由浅入深，循序渐进，读者只要具备基本的 Java 基础，就可以通过阅读本书学习 Android 应用开发的相关知识。

　　本书中的案例都是针对知识点精心设计的，并且有相应的练习题供读者练习。只要按照书中解决问题的步骤做下去，读者就会对所学知识点有一个清楚的认知，通过将书中对应的案例独立练习一遍，可以进一步巩固相应的知识点，从而做到"学中做，做中学"。

　　本书还具有以下特点。

- 基础知识全面。本书深入阐述了 Android 应用开发的核心基础组件，并且详细介绍了 Android 图形界面组件的功能和用法、Android 各种资源的管理和用法、事件处理、Android 输入和输出处理、音频和视频等多媒体开发、网络通信、Android 和 HTML5 混合开发技术等，其涉及的内容是所有 Android 开发人员必备的知识。
- 案例驱动，实用性强。本书采用"案例驱动"的方式讲解知识点，每个知识点都可以找到相应的参考案例，并且编者有针对性地设计了相应的练习题供读者独立实践，使读者能够通过实践巩固知识点。
- 通俗易懂，讲解详细。只要读者具备一定的 Java 编程基础，在阅读本书后就可以很轻松地进行 Android 应用开发。

　　本书编者具有多年从事 Java 及 Android 移动应用开发的教学经验。本书的第 1 章和第 2 章由罗佳编写，第 3 章～第 14 章由吴绍根编写。在编写过程中，企业工程师吴边等提供了大量真实的案例和宝贵的建议，在此，一并表示衷心的感谢！

　　为了方便教师教学，本书提供了案例的完整源代码等相关资源，请有此需要的教师登录华信教育资源网（www.hxedu.com.cn），在注册后免费下载，如果有问题，则可以在网站留言板留言或与电子工业出版社联系（E-mail：hxedu@phei.com.cn）。

　　由于编者水平有限，编写时间仓促，书中难免存在疏漏和不足，恳请同行专家和读者给予批评和指正。

编　者

目　录

第 1 章　Android 概述 1
1.1　Android 是什么 1
1.2　Android 应用程序的组成部分 2
1.3　Android 的发展历史 3
1.4　Android 开发环境概述 4

第 2 章　建立 Android 开发环境 5
2.1　下载和安装 Android Studio 5
2.2　开发第一个 Android 应用程序 5
2.2.1　创建 First 应用程序工程 5
2.2.2　运行 First 应用程序 7
2.3　Android 应用程序工程的结构 10
2.4　同步练习 11

第 3 章　剖析 Android 应用程序 12
3.1　AndroidManifest.xml 12
3.2　MainActivity.java——Activity 14
3.3　Android 应用程序资源 18
3.3.1　字符串资源 18
3.3.2　布局资源 19
3.3.3　id 资源 19
3.3.4　图片资源 22
3.3.5　Android 中的其他资源 24
3.3.6　引用资源 24
3.4　同步练习 26

第 4 章　深入分析 Activity 27
4.1　Activity 的生命周期 27
4.2　Activity 生命周期案例 29
4.3　使用 Log 类输出程序调试信息 32
4.4　Android 中常见的 Activity 35
4.5　同步练习 35

第 5 章 Android 中常用的 UI 组件 36

5.1 使用基于 XML 的布局 36
5.2 Android 中的基本组件 39
    5.2.1 Button 组件 39
    5.2.2 TextView 组件 42
    5.2.3 ImageView 组件 44
    5.2.4 EditText 组件 44
    5.2.5 CheckBox 组件 44
    5.2.6 RadioButton 组件 44
5.3 同步练习一 45
5.4 Android 中的容器组件 45
    5.4.1 LinearLayout 容器组件 45
    5.4.2 RelativeLayout 容器组件 50
    5.4.3 FrameLayout 容器组件 52
    5.4.4 ScrollView 容器组件 55
    5.4.5 ConstraintLayout 容器组件 60
5.5 同步练习二 63
5.6 AdapterView 组件 63
    5.6.1 AdapterView 组件入门 64
    5.6.2 Adapter 接口 64
    5.6.3 ListView 组件 65
    5.6.4 Spinner 组件 78
    5.6.5 GridView 组件 85
5.7 同步练习三 89
5.8 Android 中的其他常用组件 89
5.9 同步练习四 90

第 6 章 样式和主题 91

6.1 样式入门 91
6.2 定义样式 95
    6.2.1 定义样式的一般格式 95
    6.2.2 样式定义中的可用属性 96
6.3 应用样式 97
    6.3.1 将样式应用于某个组件上 97
    6.3.2 将样式应用于某个 Activity 或整个 Application 上 98
6.4 使用 Android 平台已定义的样式和主题 99
6.5 Android 应用程序的主题结构分析 99
6.6 同步练习 101

第 7 章 理解和使用 Intent 102

7.1 Intent 入门 102
7.2 同步练习一 106
7.3 细说 Intent 106
7.3.1 Intent 的 action 109
7.3.2 Intent 的 data 110
7.3.3 Intent 的 category 111
7.3.4 Intent 的 extra 112
7.4 Intent 解析 113
7.5 获取 Activity 返回的结果 113
7.6 Intent 的综合应用案例 119
7.6.1 运行效果 119
7.6.2 程序代码 120
7.7 同步练习二 126
7.8 广播消息和广播接收器 126
7.8.1 发送和接收普通消息 126
7.8.2 接收 Android 平台广播的普通消息 132
7.9 同步练习三 132

第 8 章 构建菜单应用程序 133

8.1 菜单 133
8.2 同步练习 137

第 9 章 动画 138

9.1 View 动画之补间动画基础 138
9.1.1 补间动画举例 138
9.1.2 补间动画的形式 141
9.1.3 使用动画监听器接口 145
9.2 View 动画之帧动画 146
9.3 同步练习 150

第 10 章 多媒体播放 151

10.1 播放音频 151
10.1.1 播放简短音频 151
10.1.2 使用 MediaPlayer 自制一个音频播放器 155
10.2 同步练习一 171
10.3 播放视频 171
10.4 同步练习二 173

第 11 章 存储程序数据 174

11.1 使用 SharedPreferences 存储程序数据 174
11.2 同步练习一 179
11.3 设置应用程序的首选项 179
11.4 同步练习二 189
11.5 在应用程序目录下存储程序数据 189
11.6 同步练习三 190
11.7 访问外部存储器 190
    11.7.1 检查 SD 卡的状态 191
    11.7.2 获取 SD 卡中特定子目录的 File 对象 191
11.8 使用 SQLite 数据库存储程序数据 192
    11.8.1 SQLite 数据库简介 192
    11.8.2 在 Android 中使用 SQLite 数据库 192

第 12 章 使用后台任务 204

12.1 使用 Java 线程执行后台任务 204
12.2 同步练习一 208
12.3 使用 AsyncTask 工具类执行后台任务 209
12.4 使用 Service 完成后台任务 215
12.5 同步练习二 223

第 13 章 使用网络 224

13.1 使用 ConnectivityManager 管理网络状态 224
13.2 使用 HttpURLConnection 访问网络 226
    13.2.1 使用 HttpURLConnection 的 GET 方法获取图片 228
    13.2.2 使用 HttpURLConnection 的 POST 方法获取图片 234
13.3 同步练习一 237
13.4 使用 OkHttp 访问网络 237
    13.4.1 使用 GET 方法进行服务请求 237
    13.4.2 使用 POST 方法进行服务请求 239
    13.4.3 构造请求头及读取响应头 241
    13.4.4 配置 OkHttp 超时 242
13.5 OkHttp GET 实现案例 242
13.6 OkHttp POST 实现案例 247
13.7 同步练习二 252
13.8 使用 Multipart 传递请求数据到服务器端 252
13.9 同步练习三 260
13.10 使用 JSON 格式的数据与服务器端通信 260
    13.10.1 JSON 基础 260

	13.10.2	在 JavaScript 中使用 JSON 数据	261
13.10.3	在 Java 中使用 JSON 数据	261
13.10.4	使用 POST 方法及 JSON 数据格式发送请求	263

第 14 章 Android 和 HTML5 的混合开发 ... 270

- 14.1 Android 和 HTML5 的混合开发基础 ... 270
- 14.2 使用 WebView 组件显示本地页面 ... 272
- 14.3 Android 与 HTML5 页面之间的信息交互 ... 274
- 14.4 同步练习 ... 284

第 1 章

Android 概述

说到智能手机，大家可能马上就能联想到 Android（安卓）、iOS 和 Windows Phone 等手机操作系统。随着 Android 手机的用户体验越来越好，应用越来越丰富，Android 智能手机已经成为智能手机的主流。因此，作为应用开发人员，开发基于 Android 的智能手机应用程序是一个重要的方向。

1.1 Android 是什么

Android 到底是什么？这个问题有点抽象，但是，即将踏入 Android 开发阵营的读者需要了解这个问题的答案。只需要了解就可以了，即使在阅读本节内容时对有些概念不太理解也是正常的，不会影响将来开发出高水平的 Android 应用。

Android 是什么？Android 是一个平台，它包括基础系统、开发工具和完整的文档。Android 平台是一个通用的计算平台，它采用 Linux 作为支撑操作系统，采用 Java 作为开发环境，通过编程实现完整的电话、视频、网络、界面设计等基础功能。Android 平台的体系结构如图 1-1 所示。下面采用从下往上的方式介绍 Android 平台体系结构的组成部分。

根据图 1-1 可知，处于 Android 平台底层的是 Linux 操作系统，在 Android 平台的体系结构中，Linux 操作系统提供了基础的支撑功能，包括设备管理、进程管理、文件管理等功能。理论上，Google 完全可以开发一套自己的操作系统，无须采用 Linux 操作系统作为 Android 平台的支撑操作系统，但是，Linux 是一个成熟、应用广泛、开源的操作系统，因此，Google 采用 Linux 操作系统作为 Android 平台的支撑操作系统。

在图 1-1 中，Linux 操作系统上方是 Core C Libraries（核心 C 语言程序库）。如前所述，Linux 是一个基础的支撑操作系统，但是在编写应用程序时，应该如何使用 Linux 操作系统提供的功能呢？Linux 操作系统为了让应用程序使用它提供的功能，专门提供了一系列 C 语言方法供应用程序调用，使应用程序可以充分利用 Linux 操作系统提供的功能，从而达到应用程序的业务目标。作为 Android 应用程序开发者，除非将来要开发具有特定功能的应用程序，在一般情况下，读者不会直接使用这些 C

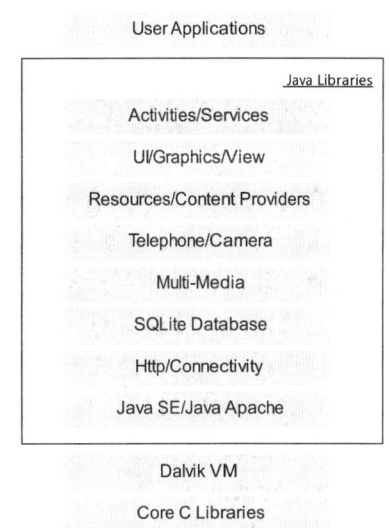

图 1-1 Android 平台的体系结构

语言方法库。在你需要使用 Core C Libraries 开发 Android 应用程序时，恭喜你，你是 Android 应用程序开发的行家了。

在图 1-1 中，Core C Libraries 上方是 Dalvik VM。Dalvik VM 是一个变形的 Java 虚拟机（Java VM），与普通的 Java 虚拟机有一点区别。Google 为什么要实现一个变形的 Java 虚拟机呢？原因很简单，因为普通的 Java 虚拟机在普通的个人计算机上可以运行得很好，可是在小型的设备（如手机）上就不一定了。因此，Google 实现了一个变形的 Java 虚拟机，并且将其命名为 Dalvik VM。为什么 Dalvik VM 出现在这里呢？因为 Android 应用开发是基于 Java 语言进行的，类似于开发桌面 Java 应用程序，我们需要一个 Java 虚拟机，用于运行开发的 Java 程序。

在图 1-1 中，Dalvik VM 上方是 Java Libraries，也就是一组 Java 类库。为什么 Java Libraries 会出现在这里？我们已经知道，Android 应用程序是基于 Java 语言开发的，为了开发 Java 应用程序，我们需要使用 Java 类库实现某些基本功能，Java Libraries 就是在编写 Android 应用程序时使用的一组 Java 类库。Java Libraries 中都包含哪些 Java 类库？对于 Java Libraries 中的大部分类库，读者可能都不认识，因为这些类库正是本书要介绍的内容，但是读者一定认识它们其中的一个——Java SE。Java SE 中包含 String、Integer、File 等基础类，读者可以像使用 Java SE 类库中的类一样使用其他类库中的类。

在图 1-1 中，Java Libraries 上方是 User Applications，也就是用户应用程序，我们开发的所有 Android 应用程序都属于 User Applications。那么，如何开发 Android 应用程序呢？这就是本书的目标。

我们从下往上介绍了 Android 平台的体系结构，现在，我们再从上往下将这些内容串起来，看看如何开发一个 Android 应用程序，以及开发的 Android 应用程序是如何在 Android 设备上运行的，具体步骤如下。在使用 Java 语言开发 Android 应用程序（User Application）时，可以使用 Java Libraries 中提供的 Java 类库实现所需的功能，在开发完成后，可以在 Dalvik VM 上运行开发的 Android 应用程序，Dalvik VM 会解析 Java 代码，并且在 Linux 操作系统上执行代码，进而完成 Android 应用程序要实现的业务功能。

1.2 Android 应用程序的组成部分

在对 Android 体系有了初步的了解后，作为 Android 应用程序开发者，读者一定想知道 Android 应用程序是由哪些部分组成的。这些概念对一个 Android 应用开发初学者来说有些难以理解，但不要太纠结于这些概念，大致了解这些概念即可。我们会尽量采用通俗易懂的语言介绍这些概念，使读者对 Android 应用程序的组成部分有一个初步的了解。

应用程序的基本组成部分包括应用程序的界面、业务功能的处理、部件之间的数据交互、数据存储。Android 应用程序也不例外，只是采用了不同的名称。具体来说，一个 Android 应用程序包括以下基本组成部分。

1. Activity

在 Android 应用程序中，一个界面就是一个 Activity（窗体）。Activity 是界面，因此在设计 Activity 时，都会对界面的布局（Android 提供了丰富的 UI 组件，用于实现绚丽的界面）和界面组件的点击操作进行相应的设计。可以将 Android 的 Activity 类比为 Internet 的页面。

2. View

View（窗体组件）是构建应用程序界面的基本组件，也就是说，Activity 是由一个或多个

View 构成的，Button、Label、Text Field 等都是 View，View 是构建 Activity 的基本元素。

3．Intent

在一般情况下，一个 Android 应用程序中包含多个界面，用户在进行不同的操作时，可能会在不同界面之间进行切换，就像在 Internet 页面中，当用户点击不同的页面链接时，会在不同页面之间进行切换一样。Internet 页面之间的切换是通过超链接实现的，在 Android 应用程序中，Activity 之间的切换是通过 Intent（窗体之间或应用程序之间的通信组件）对象实现的。因此，在 Android 应用程序中，Intent 是界面之间及功能部件之间实现信息交互的桥梁。

4．Content Provider

Content Provider（内容提供者）是 Android 建议的应用程序之间进行数据交互的方式。如果一个 Android 应用程序希望将自己的数据提供给其他 Android 应用程序使用，那么该 Android 应用程序需要实现 Content Provider 接口，这样，其他的 Android 应用程序就可以通过这个接口访问这个 Android 应用程序提供的数据了。一个典型的实现了 Content Provider 接口的 Android 应用程序是通讯录应用程序，任何需要使用通讯录数据的 Android 应用程序都可以通过该接口从通讯录应用程序中获取通讯录数据。

5．Service

Service 是运行于后台的程序。在一般情况下，Service 没有用户界面，它运行于后台，并且为运行于前端的程序提供服务。Android 中的 Service 在运行方式上类似于 Windows Phone 中的后台进程：它们都在安静地运行，并且在需要时为其他程序提供服务。

6．广播接收器

广播接收器，即 Broadcast Receiver。Android 平台上的程序在运行时可能会发生任何事件，有些程序在运行时可能将它的事件广播出来，其他程序可以监听这样的事件，并且对发生的事件进行必要的响应。举例来说，Android 的电池电量监视程序（Android 的一个 Service）在随时监视着电池的电量，当电池的电量低于某个门槛值时，该程序会广播一条消息，其他程序可以监听这条消息，并且对这个事件进行必要的响应。例如，当一个正在进行高耗电运算的程序监听到这条消息时，应该停止高耗电的运算，以便减少对电量的消耗。

7．AndroidManifest.xml 文件

AndroidManifest.xml 文件（应用程序描述文件）是 Android 应用程序的配置文件，它将构成 Android 应用程序的各个组件有效地装配起来，从而构成一个完整的Android 应用程序。每个应用程序都包含一个且只包含一个 AndroidManifest.xml 配置文件。

1.3 Android 的发展历史

2007 年，Google 建立了开放手机联盟（Open Handset Alliance）。2009 年，这个联盟成员包括 Sprint Nextel、T-Mobile、Motorola、Samsung、Sony Ericsson、Toshiba、Vodafone、Google、Intel、Texas Instruments 等 IT 巨头。2011 年，开放手机联盟的成员已近 80 家，Android 已经成为移动设备事实上的行业标准。2022 年 3 月，Android 平台的版本已从 1.0 发展到了 12.0。Android 发展迅速，因此，读者必须学会使用 Android 的在线帮助文档。

1.4　Android 开发环境概述

　　Android 平台使用 Java 作为应用程序开发语言。Android 开发环境是指 Android 应用程序的开发环境，包括 Java 基本包、Android 基础组件、Android UI 组件、Android 服务组件、Android 电话和媒体服务组件、Android 模拟器（Android Virtual Device，AVD）、Android 调试器等。

　　编者将在后续章节中对这些内容进行详细介绍。首先从建立 Android 开发环境开始讲解。

第 2 章 建立 Android 开发环境

Google 为开发 Android 应用程序提供了完整的开发环境 Android Studio。在编写本书时，Android Studio 的最新版本是 Android Studio 2021.1.1 Patch 2。由于 Android App 是基于 Java 语言开发的，因此，在安装 Android Studio 前，需要安装 Java SDK，建议安装 Java SDK 8 或更高版本。

2.1 下载和安装 Android Studio

在 Android Developer 网站中下载 Android Studio 开发包，直接运行下载的安装软件，即可根据提示安装 Android Studio。在安装 Android Studio 的过程中，由于需要在线下载并构建 Android 应用程序的相关工具，因此需要计算机保持在线。在一般情况下，这个过程需要持续大约 10 分钟。

2.2 开发第一个 Android 应用程序

2.2.1 创建 First 应用程序工程

在安装 Android 开发环境后，就可以开发 Android 应用程序了。启动 Android Studio，进入 Android Studio 的启动界面，如图 2-1 所示。

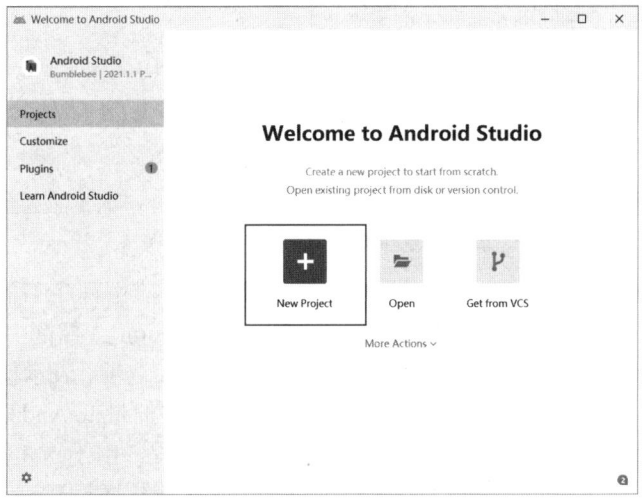

图 2-1　Android Studio 的启动界面

在 Android Studio 的启动界面中单击 New Project 按钮，可以新建一个 Android 应用程序工程，界面如图 2-2 所示。

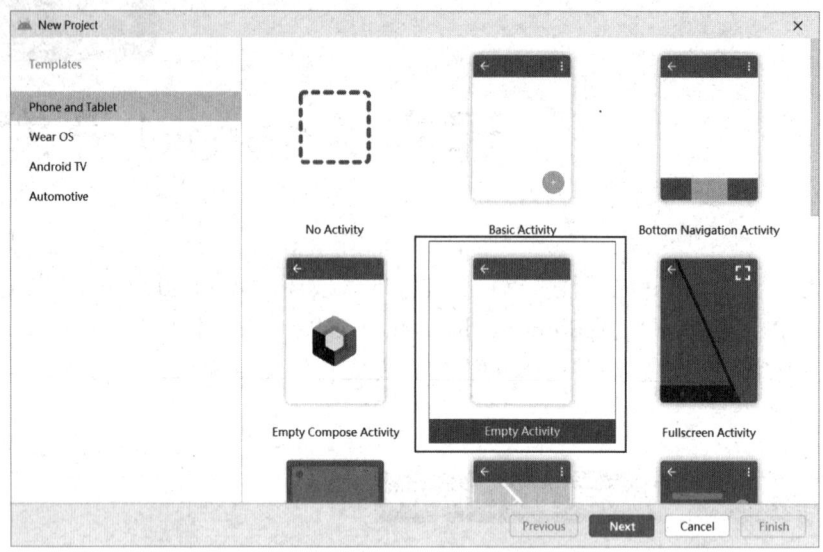

图 2-2　新建 Android 应用程序工程的界面（一）

在图 2-2 所示的界面中单击 Empty Activity 按钮，然后单击 Next 按钮，进入新建 Android 应用程序工程的界面，如图 2-3 所示。

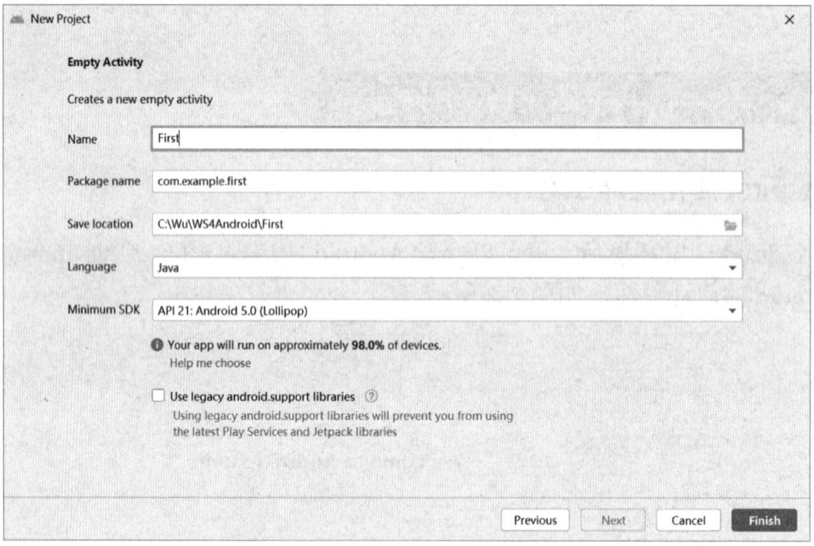

图 2-3　新建 Android 应用程序工程的界面（二）

在图 2-3 所示的界面中，在 Name 文本框中输入 Android 应用程序工程的名称，在 Package name 文本框中输入 Android 应用程序包的名字；在 Save location 选择框中输入或选择 Android 应用程序工程的存储目录；在 Language 下拉列表中选择 Java 选项，表示使用 Java 作为开发语言；在 Minimum SDK 下拉列表中选择运行 Android 应用程序的最低 Android 版本，单击 Finish 按钮，Android Studio 会创建一个新的 Android 应用程序工程。由于这是第一个 Android 应用程序工程，因此 Android Studio 需要下载所需的 Android 应用程序构建工具，如 Gradle 工具，在一般情况下，这个过程需要大约 10 分钟，完成后的界面如图 2-4 所示。

第 2 章　建立 Android 开发环境

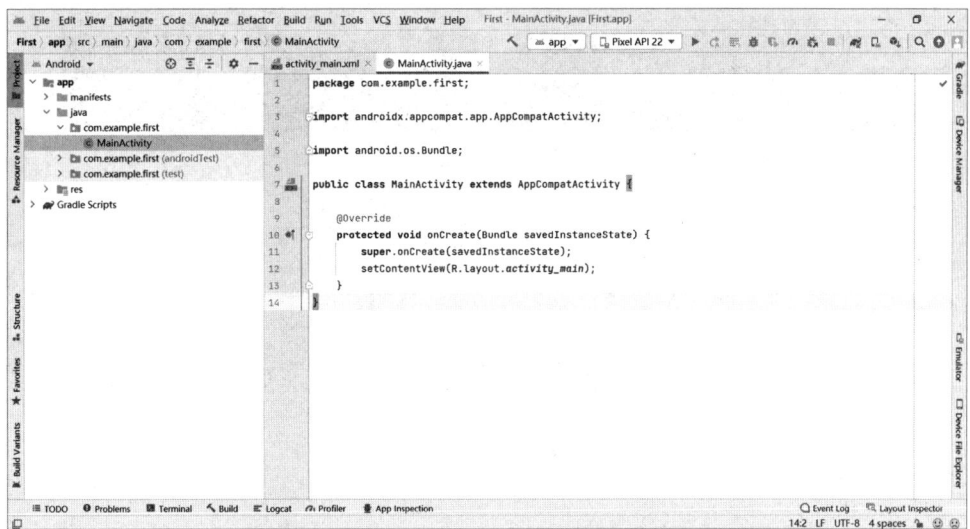

图 2-4　新建完成的 Android 应用程序工程界面

至此，Android Studio 已经成功创建了名为 First 的 Android 应用程序工程，现在可以运行这个简单的 Android 应用程序了。

2.2.2　运行 First 应用程序

Android Studio 提供了两种测试 Android 应用程序的方法，一种是直接在 Android 真机上运行，另一种是在 Android SDK 自带的模拟器（简称 Android 模拟器）上运行。我们在 Android 模拟器上测试。为此，我们需要配置一个 Android 模拟器。

在 Android 应用程序工程界面中单击界面右边的 Device Manager 按钮，进入创建 Android 模拟器的界面，如图 2-5 所示。

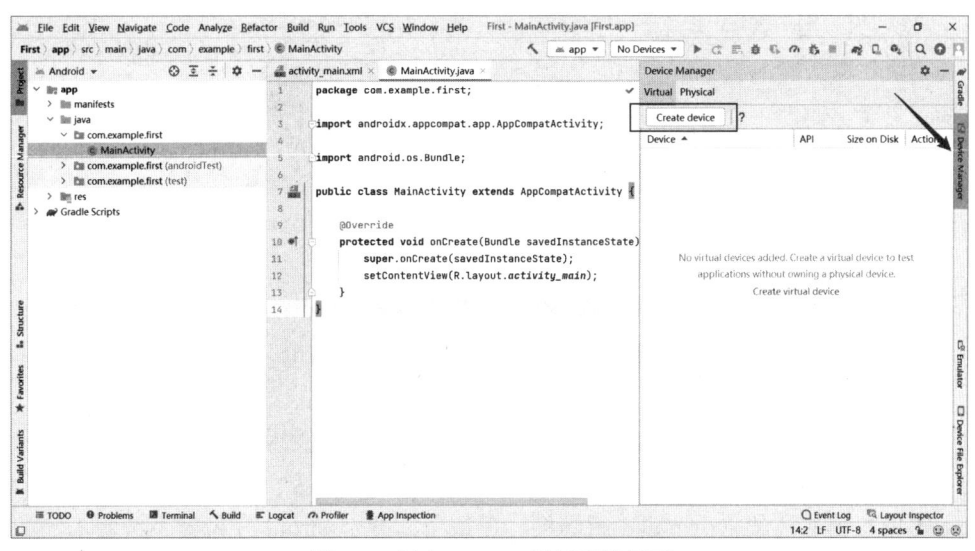

图 2-5　创建 Android 模拟器的界面

在图 2-5 所示的界面中单击 Create device 按钮，进入选择 Android 模拟器的界面，如图 2-6 所示。

7

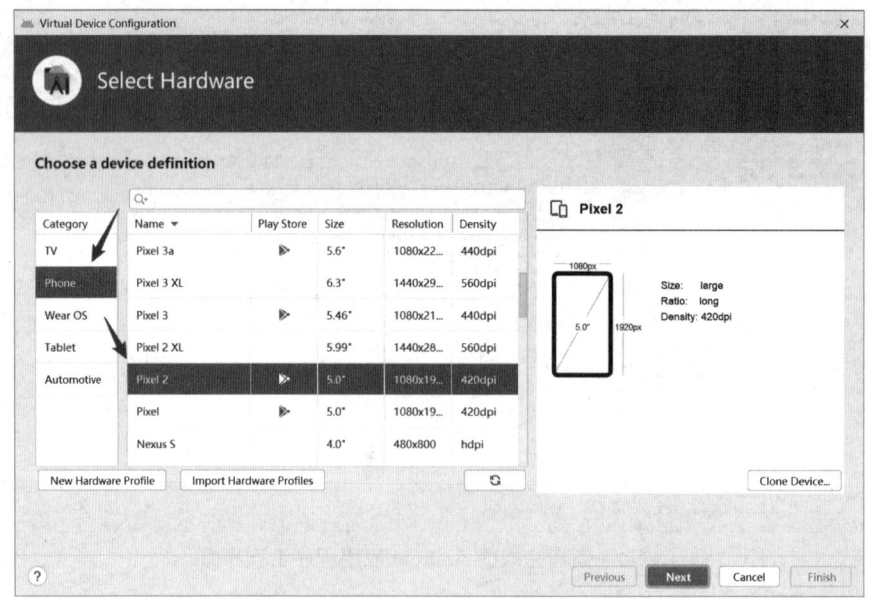

图 2-6　选择 Android 模拟器的界面

在图 2-6 所示的界面中，在第一个列表框中选择 Phone 选项，在第二个列表框中选择 Name 为 Pixel 2、Size 为 5.0 英寸的选项，单击 Next 按钮，进入选择 Android 模拟器系统版本的界面，如图 2-7 所示。

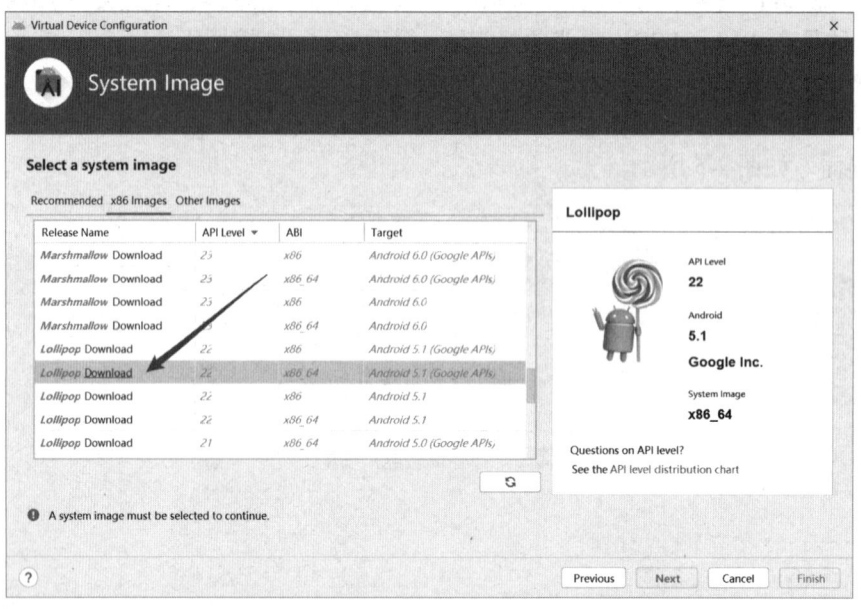

图 2-7　选择 Android 模拟器系统版本的界面

由于所安装 Android Studio 还没有安装可以模拟 Android 5.1 的模拟器，因此需要单击所选 Android 模拟器右边的 Download 超链接，下载所需的 Android 模拟器系统版本，如图 2-8 所示。在下载并安装 Android 模拟器系统后，返回图 2-7 所示的界面，单击 Next 按钮，进入完成 Android 模拟器配置界面，如图 2-9 所示。

在图 2-9 所示的界面中单击 Finish 按钮，完成 Android 模拟器的配置。此时，Android Studio 的开发界面中会显示所配置的 Android 模拟器列表，如图 2-10 所示。

第 2 章 建立 Android 开发环境

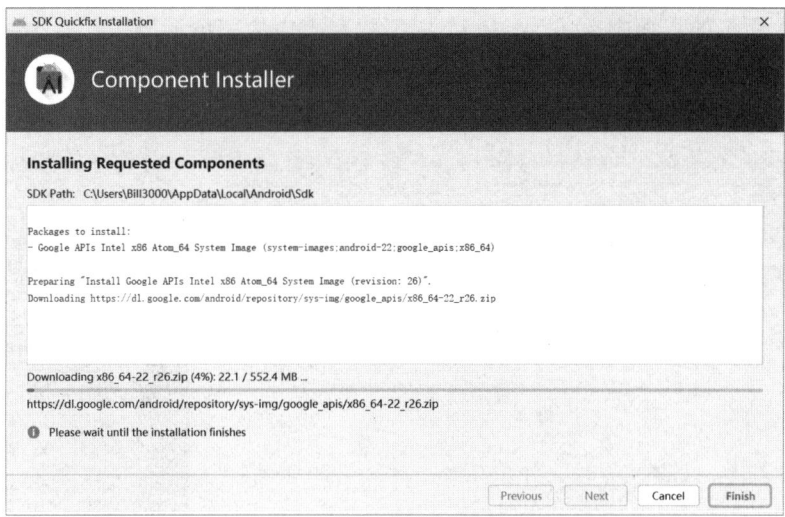

图 2-8　下载所需的 Android 模拟器系统版本

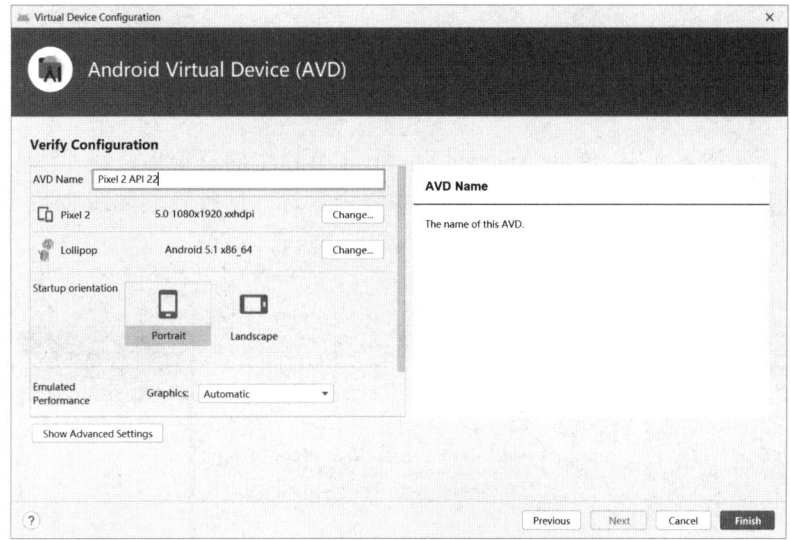

图 2-9　完成 Android 模拟器配置界面

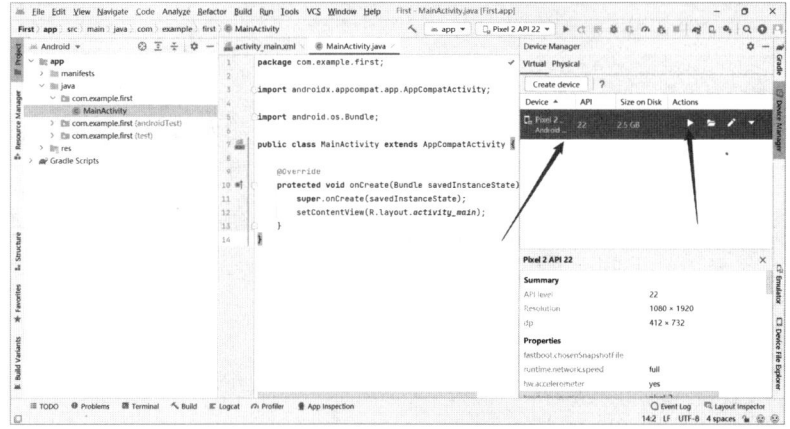

图 2-10　Android 模拟器列表

在图 2-10 所示的界面中单击 Android 模拟器列表中的三角形按钮，启动并运行 Android 模拟器，界面如图 2-11 所示。为了使 Android 模拟器有更多的屏幕显示空间，可以单击 Device Manager 按钮，收起 Device Manager 界面。

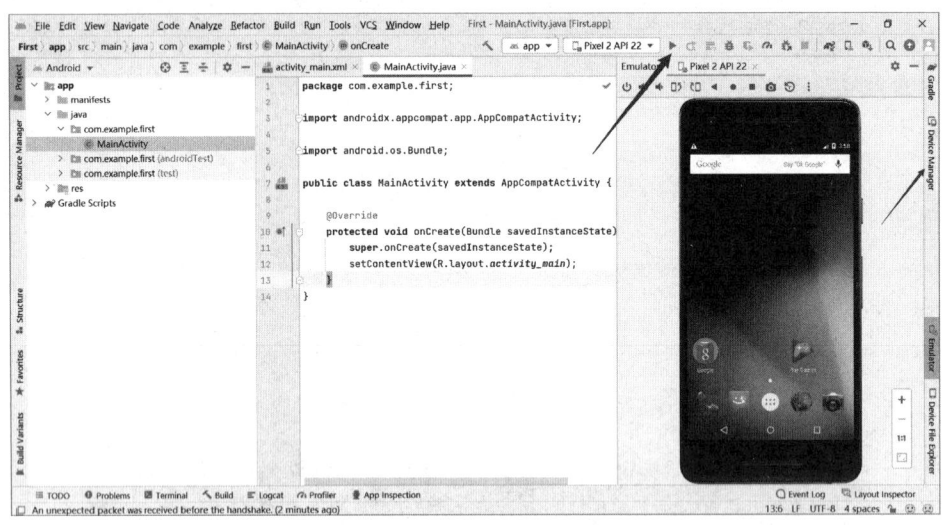

图 2-11　启动并运行 Android 模拟器后的界面

在图 2-11 所示的界面中，单击 Android Studio 工具栏中的三角形按钮，启动并运行所创建的 Android 应用程序 First，使其在 Android 模拟器上运行，运行界面如图 2-12 所示。

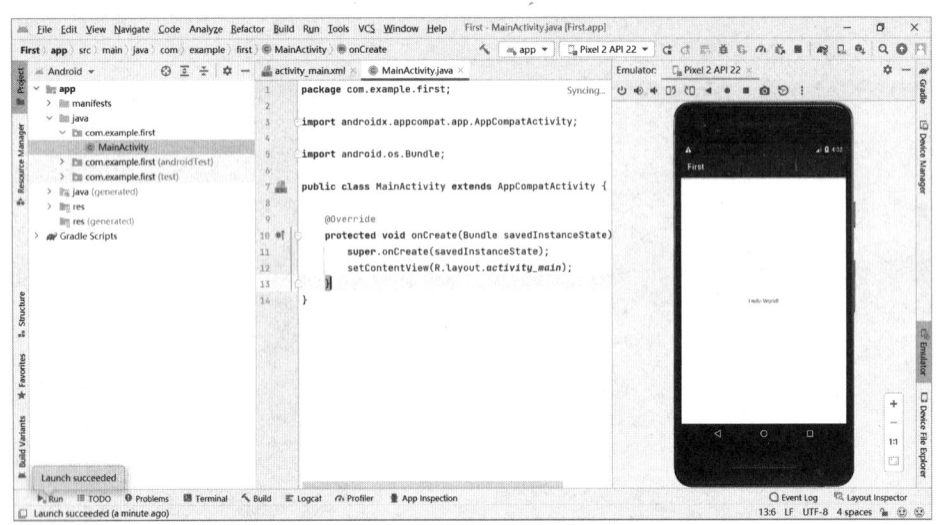

图 2-12　Android 应用程序 First 的运行界面

Android 模拟器中会显示"Hello World"。至此，我们的第一个 Android 应用程序已经成功在模拟器上运行起来了。下面，我们对 Android 应用程序工程的结构进行简单的分析。

2.3　Android 应用程序工程的结构

在不同的 Android 应用环境中，Android 应用程序工程的规模相差很大，但是，Android 应用程序工程的结构是相似的。典型的 Android 应用程序工程的结构如图 2-13 所示。

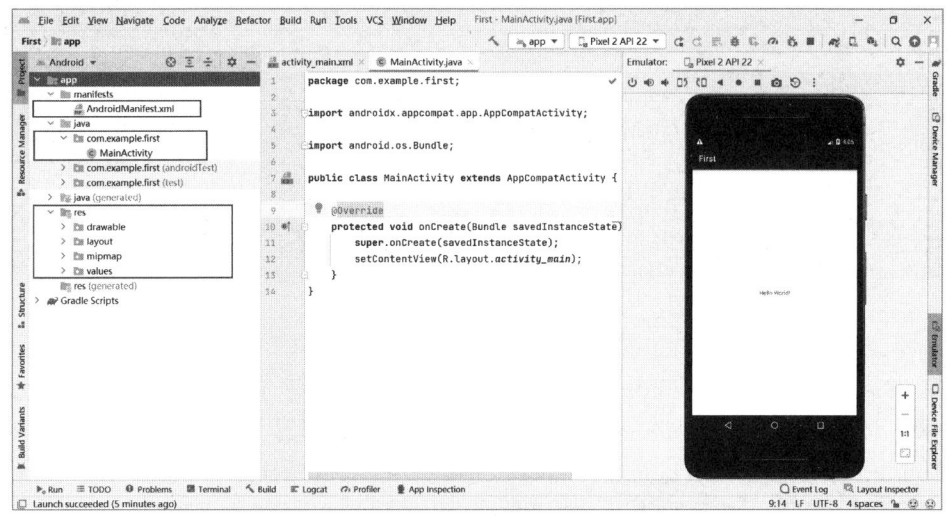

图 2-13　典型的 Android 应用程序工程的结构

Android 应用程序工程的结构的相关说明如下。
- manifests 目录下的 AndroidManifest.xml 文件。AndroidManifest.xml 文件是 Android 应用程序的配置文件，它类似于 Java EE 程序中的 web.xml 文件，该文件中包含 Android 应用程序的基本信息。例如，一个 Android 应用程序是由哪些 Activity 组成的，哪个是入口 Activity，运行该 Android 应用程序的 Android 平台最低版本，等等。
- java 目录。java 目录是应用程序的 Java 源代码程序包和源代码所在的位置，如 First 应用程序的 com.example.first 包及源代码文件 MainActivity.java 都存储于该目录下。
- res 目录。res 目录下包括 drawable、layout、values 等子目录，该目录与 java 目录一样重要，是经常需要操作的目录。res 目录主要用于存储 Android 应用程序的资源。我们将在后续章节中对这个目录下的资源类型及其格式进行详细介绍。

至此，我们已经对 Android 平台及 Android 开发环境进行了初步介绍，并且开发了第一个 Android 应用程序。接下来，我们将对 Android 应用程序进行剖析，带领读者逐步进入 Android 的精彩世界。

2.4　同步练习

建立 Android 开发环境，编写第一个 Android 应用程序工程，并且运行该应用程序，观察该应用程序工程的结构，熟悉 Android Studio 开发环境的各个功能的使用方法。

第 3 章

剖析 Android 应用程序

在第 2 章中,我们对 Android 应用程序工程的结构进行了简单的介绍,并且指出了 AndroidManifest.xml 文件是 Android 应用程序中非常重要的配置文件,在启动 Android 应用程序时,Android 平台会先读取这个文件并对其进行分析,再启动特定的 Activity,用于运行该 Android 应用程序。因此,我们从该文件开始对 Android 应用程序进行剖析。

3.1 AndroidManifest.xml

双击打开第 2 章介绍的 Android 应用程序工程 First 中的 AndroidManifest.xml 文件,可以看到以下内容。

```xml
<?xml version="1.0" encoding="utf-8"?>
<manifest xmlns:android="http://schemas.android.com/apk/res/android"
    package="com.example.first">

    <application
        android:allowBackup="true"
        android:icon="@mipmap/ic_launcher"
        android:label="@string/app_name"
        android:roundIcon="@mipmap/ic_launcher_round"
        android:supportsRtl="true"
        android:theme="@style/Theme.First">
        <activity
            android:name=".MainActivity"
            android:exported="true">
            <intent-filter>
                <action android:name="android.intent.action.MAIN" />

                <category android:name="android.intent.category.LAUNCHER" />
            </intent-filter>
        </activity>
    </application>

</manifest>
```

第 3 章 剖析 Android 应用程序

前面介绍过，AndroidManifest.xml 文件中包含 Android 应用程序的基本信息，以下代码说明 AndroidManifest.xml 文件是一个 XML 文件。

```
<?xml version="1.0" encoding="utf-8"?>
```

\<manifest\>标签如下：

```
<manifest xmlns:android="http://schemas.android.com/apk/res/android"
    package="com.example.first">
```

在\<manifest\>标签中，使用 xmlns:android 属性定义 android 前缀的名字空间，用于避免名字空间发生冲突；使用 package 属性定义应用程序的包名，需要注意的是，每个 Android 应用程序都有一个唯一的包名，这个唯一的包名就是使用\<manifest\>标签的 package 属性定义的，也是我们在创建 Android 应用程序工程时定义的包名，我们称这个包为应用的包。

\<application\>标签如下：

```
<application
    android:allowBackup="true"
    android:icon="@mipmap/ic_launcher"
    android:label="@string/app_name"
    android:roundIcon="@mipmap/ic_launcher_round"
    android:supportsRtl="true"
    android:theme="@style/Theme.First">
```

\<application\>标签主要用于定义一个 Android 应用程序，下面重点介绍 android:icon 属性和 android:label 属性。这两个属性分别用于定义 Android 应用程序在 Android 手机的应用程序管理界面中显示的图标和名称，在 Android 手机中，可以通过 Settings→Apps 命令进入应用程序管理界面，如图 3-1 所示。

图 3-1 中箭头所指的部分是 Android 应用程序的图标和名称。Android 应用程序中的图标、字符串常量等都称为 Android 应用程序资源，如@mipmap/ic_launcher 和@string/app_name，它们都是在工程的 res 目录下定义的，如图 3-2 所示。

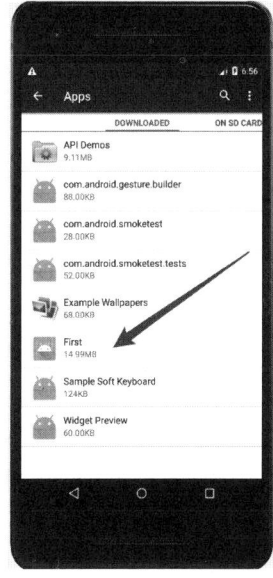

图 3-1　Android 手机的应用程序管理界面

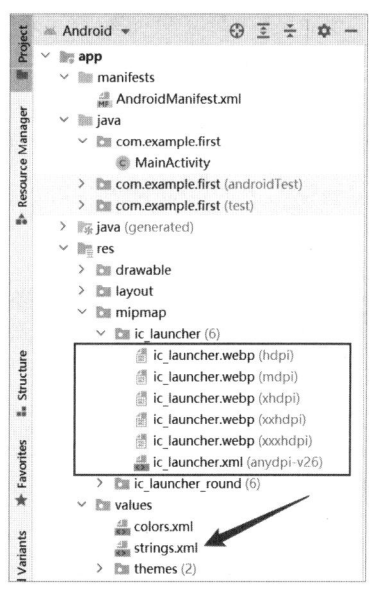

图 3-2　Android 应用程序资源

打开 mipmap 目录下的 ic_launcher.webp 文件，会显示一个小机器人图标，这个图标就是显示在应用列表中的应用程序图标。打开 values 目录下的 strings.xml 文件，可以看到以下内容。

```
<resources>
    <string name="app_name">First</string>
</resources>
```

其中的 app_name 对应于@string/app_name 指定的字符串资源。

继续观察 AndroidManifest.xml 文件中的内容，接下来的内容如下：

```
<activity
    android:name=".MainActivity"
    android:exported="true">
```

这段 XML 代码定义了 Android 应用程序的一个 Activity，即 Android 应用程序的一个界面。Android 应用程序的一个 Activity 相当于网站的一个页面，这个 Activity 的主类由 android:name 属性定义，<manifest>标签的 package 属性值加上 android:name 属性值就构成了 Activity 的主类，因此，这个 Activity 的主类就是 com.example.first.MainActivity。android:exported 属性主要用于设置这个 Activity 是否可以被其他 Android 应用程序打开，如果其值为 true，则表示其他的 Android 应用程序可以通过 Intent 打开这个 Activity。

继续观察 AndroidManifest.xml 文件中的内容，接下来的内容如下：

```
<intent-filter>
    <action android:name="android.intent.action.MAIN" />

    <category android:name="android.intent.category.LAUNCHER" />
</intent-filter>
```

这段 XML 代码定义了入口 Activity。下面介绍如何定义入口 Activity。Android 应用程序中有一个称为入口 Activity 的 Activity，该 Activity 是 Android 应用程序的入口。在启动 Android 应用程序时，会首先运行入口 Activity。入口 Activity 类似于 C 语言程序的 main()方法，网站程序的 index.html 文件，Java 桌面程序的 main()方法。那么，如何将一个 Activity 定义为入口 Activity 呢？通过<intent-filter>标签定义。在<intent-filter>标签中，通过将<action>标签中的 android:name 属性值设置为 android.intent.action.MAIN，以及将<category>标签中的 android:name 属性值设置为 android.intent.category.LAUNCHER，将一个 Activity 定义为入口 Activity？

至此，我们对 AndroidManifest.xml 文件进行了详细的剖析，下面对 MainActivity.java 文件中的代码进行剖析。

3.2 MainActivity.java——Activity

在启动 Android 应用程序时，Android 平台会先读取 AndroidManifest.xml 文件，从中获取入口 Activity 的相关信息，再启动入口 Activity，用于运行该应用程序。在 Android 应用程序 First 中，由于入口 Activity 是 MainActivity，因此，Android 平台会首先启动并运行 MainActivity，从而显示应用程序界面。MainActivity.java 文件中的代码如下：

```
package com.example.first;
```

```java
import androidx.appcompat.app.AppCompatActivity;

import android.os.Bundle;

public class MainActivity extends AppCompatActivity {

    @Override
    protected void onCreate(Bundle savedInstanceState) {
        super.onCreate(savedInstanceState);
        setContentView(R.layout.activity_main);
    }
}
```

这段代码很简单。当 Android 平台启动并运行 MainActivity 时，系统会首先调用 onCreate() 方法。需要注意的是，Android 要求，必须先调用父类的 onCreate() 方法，再进行自己的初始化工作。在 MainActivity 文件中的 onCreate() 方法中，在调用了父类的 onCreate() 方法后，调用 setContentView(R.layout.activity_main) 方法，用于显示 Activity 的主界面。

在 Android 应用程序中，有两种构建应用程序界面的方法，一种是使用布局资源（Layout Resource）文件（简称布局文件）构建界面，另一种是使用程序代码构建界面。我们使用布局文件构建 Android 应用程序界面，这也是常用的构建 Android 应用程序界面的方法。

注意以下代码。

```java
setContentView(R.layout.activity_main);
```

其中，参数 R.layout.activity_main 就是界面布局文件，它对应工程的 res/layout 目录下的 activity_main.xml 文件，如图 3-3 所示。

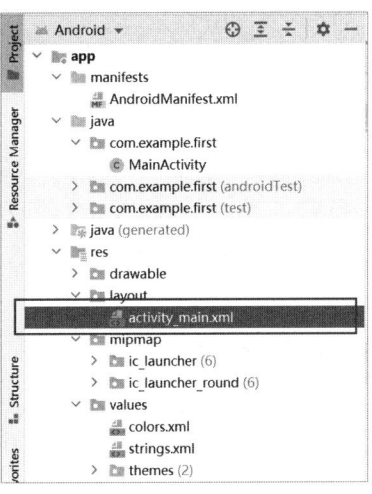

图 3-3　activity_main.xml 文件

读者可以这样理解 R.layout.activity_main：R 代表 res 目录，layout 代表 res 目录下的 layout 目录，activity_main 代表 res/layout 目录下的 activity_main.xml 文件。打开 activity_main.xml 文件，Android Studio 会显示如图 3-4 所示的界面。

Android Studio 移动应用开发基础（第 2 版）

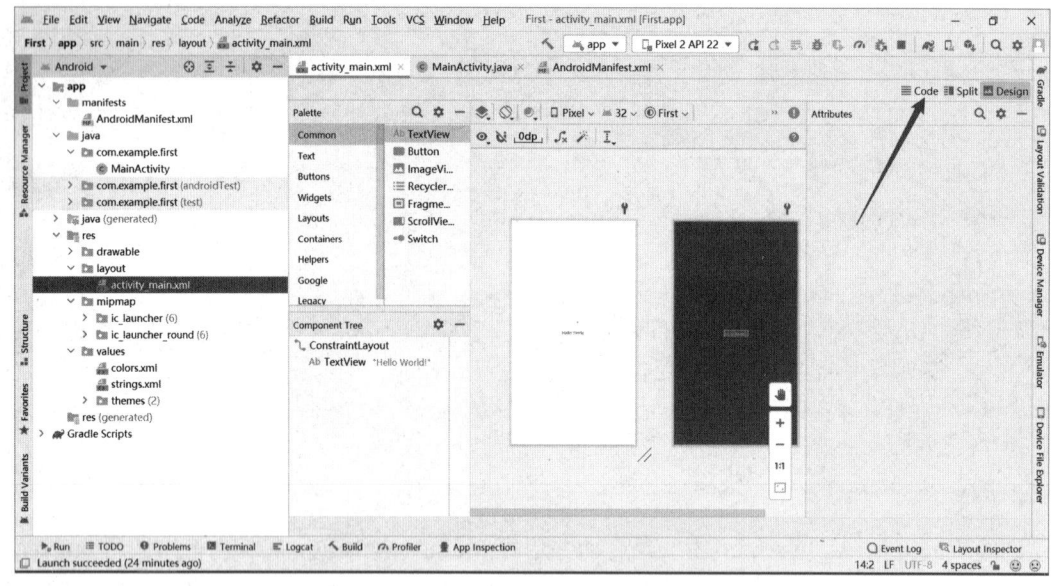

图 3-4　打开 activity_main.xml 文件时的界面

Android Studio 提供了图形化的界面设计工具。由于 activity_main.xml 文件是 Android 应用程序的布局文件，因此，当打开这个文件时，Android Studio 会自动打开界面的图形化设计工具。单击图 3-4 中箭头所指的 Code 按钮，切换到界面布局的代码编写模式，即可看到 activity_main.xml 文件中的代码，具体如下：

```xml
<?xml version="1.0" encoding="utf-8"?>
<androidx.constraintlayout.widget.ConstraintLayout xmlns:android="http://schemas.android.com/apk/res/android"
    xmlns:app="http://schemas.android.com/apk/res-auto"
    xmlns:tools="http://schemas.android.com/tools"
    android:layout_width="match_parent"
    android:layout_height="match_parent"
    tools:context=".MainActivity">

    <TextView
        android:layout_width="wrap_content"
        android:layout_height="wrap_content"
        android:text="Hello World!"
        app:layout_constraintBottom_toBottomOf="parent"
        app:layout_constraintLeft_toLeftOf="parent"
        app:layout_constraintRight_toRightOf="parent"
        app:layout_constraintTop_toTopOf="parent" />

</androidx.constraintlayout.widget.ConstraintLayout>
```

根据上述代码可知，使用 ConstraintLayout 容器组件对组件的布局进行管理，这个容器组件在水平方向上会占满整个屏幕，代码为 android:layout_width="match_parent"，在垂直方向上也会占满整个屏幕，代码为 android:layout_height="match_parent"。ConstraintLayout 容器组件

第 3 章 剖析 Android 应用程序

中包含一个称为 TextView 的可视组件，这个组件可以显示一行文本信息，在水平方向上，这个组件占用的空间足够显示其内部的文字即可，这就是 android:layout_height= "wrap_content" 和 android: layout_width="wrap_content"的含义。

至此，我们可以很容易地理解下面的代码。

```
setContentView(R.layout.activity_main);
```

这条代码的作用是将 R.layout.activity_main 指定的界面显示在 MainActivity 的主窗口中。

读者可以根据前面介绍的知识做一道练习题：将显示在屏幕上的"Hello World"改为"你好，世界!"。然后在"你好，世界!"的下面显示"Android 开发入门了!"。

在 Android 应用程序中，我们可以在任何地方直接使用字符串常量，就像在 TextView 组件中使用的那样。但是，在 Android 中不建议这么做，Android 建议将字符串常量放到 res/values/strings.xml 文件中进行集中管理。为此，我们将 res/layout/activity_main.xml 文件中的代码修改为以下代码（读者课堂练习的结果）。

```xml
<?xml version="1.0" encoding="utf-8"?>
<androidx.constraintlayout.widget.ConstraintLayout
    xmlns:android="http://schemas.android.com/apk/res/android"
    xmlns:app="http://schemas.android.com/apk/res-auto"
    xmlns:tools="http://schemas.android.com/tools"
    android:layout_width="match_parent"
    android:layout_height="match_parent"
    tools:context=".MainActivity">

    <TextView
        android:id="@+id/tv01"
        android:layout_width="wrap_content"
        android:layout_height="wrap_content"
        android:text="@string/hello"
        app:layout_constraintTop_toTopOf="parent"
        app:layout_constraintLeft_toLeftOf="parent"
        app:layout_constraintRight_toRightOf="parent" />

    <TextView
        android:layout_width="wrap_content"
        android:layout_height="wrap_content"
        android:text="@string/started"
        app:layout_constraintTop_toBottomOf="@id/tv01"
        app:layout_constraintLeft_toLeftOf="parent"
        app:layout_constraintRight_toRightOf="parent" />

</androidx.constraintlayout.widget.ConstraintLayout>
```

在上述代码中，同样使用 ConstraintLayout 容器组件对组件的布局进行管理。ConstraintLayout 容器组件中包含两个 TextView 组件，并且分别为这两个 TextView 组件分配了可以唯一标识它们的 id：tv01 和 tv02。在第一个 TextView 组件的布局属性中，通过代码"app:layout_constraintTop_ toTopOf="parent"" 将第一个 TextView 组件放置在 ConstraintLayout 容器组件的顶部；在第二个 TextView 组件的布局属性中，通过代码 "app:layout_constraintTop_toBottomOf="@id/tv01"" 将第二个 TextView 组件放置在第一个组件的下面。此外，为了将字符串常量与布局文件分离，需要在 res/values/strings.xml 文件中增加对 hello 和 started 字符串资源的定义，代码如下：

```xml
<resources>
    <string name="app_name">First</string>
    <string name="hello">你好，世界！</string>
    <string name="started">Android开发入门了！</string>
</resources>
```

将 Android 应用程序资源与 Android 应用程序代码分离，可以更好地对 Android 应用程序进行后期维护。例如，如果要使应用程序界面显示德文，那么只需修改资源文件，无须修改程序源代码。综上所述，可以得出以下结论。

Android 应用程序 = Java 程序代码文件 + 资源文件 + AndroidManifest.xml 文件

粗略地讲，以上结论是正确的。

3.3 Android 应用程序资源

Android 应用程序资源在 Android 应用程序中起着十分重要的作用。Android 应用程序资源可以是一个文件，如布局文件；也可以是一个值，如字符串常量。将 Android 应用程序资源与 Android 应用程序代码分离的好处是，可以直接改变资源的值，而不用修改或编译 Android 应用程序代码。

在 Android 应用程序中，会用到各种各样的资源，包括字符串资源、布局资源、id 资源、图片资源、动画资源等，下面我们对常用的资源类型进行简单的介绍。

3.3.1 字符串资源

在程序编码实践中，我们经常会用到大量的字符串常量。因此，Android 建议将字符串常量统一定义到一个或多个 XML 资源文件中。例如，在 First 应用程序工程中，我们将用到的字符串常量统一定义到 res/values/strings.xml 文件中。需要注意的是，用于定义字符串常量的 XML 文件必须存储于工程的 res/values 目录下，而文件名可以根据需要自行定义。例如，将 strings.xml 文件重命名为 constants.xml，Android 应用程序照样可以正常运行。根据需要，可以将字符串常量定义到多个文件中。例如，将原来存储于 strings.xml 文件中的内容拆分到文件 constants.xml 和 another.xml 中，只要这两个文件都存储于 res/values 目录下，Android 应用程序就可以正常运行。

constants.xml 文件中的代码如下：

```xml
<resources>
    <string name="app_name">HelloWorld</string>
    <string name="hello">你好，世界！</string>
</resources>
```

another.xml 文件中的代码如下：

```xml
<resources>
    <string name="started">Android开发入门了!</string>
</resources>
```

运行修改后的 Android 应用程序，会得到与前面完全一样的结果。

Android Studio 的构建工具会给每个资源都分配一个唯一的编号，通过这个编号可以访问特定的资源。例如，为了在 Java 代码中访问 res/values/strings.xml 文件中定义的字符串资源，可以使用以下代码访问由 app_name 属性定义的字符串资源。

```
R.string.app_name
```

在 XML 资源配置文件（如 AndroidManifest.xml 文件）中，可以使用以下代码访问由 app_name 属性定义的字符串资源。

```
@string/app_name
```

3.3.2 布局资源

在 Android 中，可以使用 Java 代码直接构建应用程序界面，但通常使用 XML 文件构建应用程序界面。构建应用程序界面的文件称为布局文件。例如，在 First 应用程序工程中，res/layout/activity_main.xml 文件就是一个布局文件。在创建该布局文件后，可以通过程序代码直接将该界面显示在 Activity 中，代码如下：

```java
public class MainActivity extends Activity {

    @Override
    protected void onCreate(Bundle savedInstanceState) {
        super.onCreate(savedInstanceState);
        setContentView(R.layout.activity_main);    //显示布局界面
    }
    ...
}
```

Android 规定，所有的布局文件都必须存储于 res/layout 目录下。因此，如果有多个界面需要显示，那么在 res/layout 目录下会有多个布局文件。

注意以下代码。

```
setContentView(R.layout.activity_main);
```

类似于在 Java 代码中引用字符串资源，在 Java 代码中，可以使用 R.layout.activity_main 布局文件引用特定的布局资源。

现在对 Android 应用程序的功能稍做改进：在运行 Android 应用程序时，将第二个 TextView 组件中显示的字符串"Android 开发入门了！"动态改为"欢迎进入 Android 阵营！"，应该如何做呢？我们需要在 Android 应用程序的代码中获取第二个 TextView 组件的引用，然后通过 TextView 对象提供的方法修改其中显示的字符串。为了能够在 Java 代码中获取布局资源中某个对象的引用，我们需要给所引用的组件赋予一个 id，这就是下一节要介绍的内容。

3.3.3 id 资源

就像给字符串资源、界面布局资源定义唯一的标识符一样，我们也可以给界面布局资源中的每个组件都分配一个唯一的 id。在给界面布局资源中的组件分配唯一的 id 后，就可以在 Android 应用程序的代码中引用这些组件了。因此，我们需要在界面布局资源中为每个需要指定 id 的组件加上 android:id 属性。例如，在 res/layout/activity_main.xml 文件中，可以给第二个 TextView 组件赋予一个 id，代码如下：

```xml
<?xml version="1.0" encoding="utf-8"?>
<androidx.constraintlayout.widget.ConstraintLayout
    xmlns:android="http://schemas.android.com/apk/res/android"
    xmlns:app="http://schemas.android.com/apk/res-auto"
    xmlns:tools="http://schemas.android.com/tools"
    android:layout_width="match_parent"
    android:layout_height="match_parent"
    tools:context=".MainActivity">

    <TextView
        android:id="@+id/tv01"
        android:layout_width="wrap_content"
        android:layout_height="wrap_content"
        android:text="@string/hello"
        app:layout_constraintTop_toTopOf="parent"
        app:layout_constraintLeft_toLeftOf="parent"
        app:layout_constraintRight_toRightOf="parent" />

    <TextView
        android:id="@+id/tv02"
        android:layout_width="wrap_content"
        android:layout_height="wrap_content"
        android:text="@string/started"
        app:layout_constraintTop_toBottomOf="@id/tv01"
        app:layout_constraintLeft_toLeftOf="parent"
        app:layout_constraintRight_toRightOf="parent" />

</androidx.constraintlayout.widget.ConstraintLayout>
```

在上面的代码中，给第一个 TextView 组件分配名为 tv01 的 id；给第二个 TextView 组件分配名为 tv02 的 id。在 Java 程序代码中，可以通过以下代码获取 id 为 tv02 的组件的引用。

```java
TextView tv = this.findViewById(R.id.tv02);
```

然后通过以下代码修改显示在组件中的内容。

```java
tv.setText("欢迎进入Android阵营！");
```

修改后的 MainActivity.java 文件中的完整代码如下：

```java
package com.example.first;

import androidx.appcompat.app.AppCompatActivity;
```

```java
import android.os.Bundle;
import android.widget.TextView;

public class MainActivity extends AppCompatActivity {

    @Override
    protected void onCreate(Bundle savedInstanceState) {
        super.onCreate(savedInstanceState);
        setContentView(R.layout.activity_main);
        TextView tv = findViewById(R.id.tv02);
        tv.setText("欢迎进入Android阵营！");
    }
}
```

运行修改后的 First 应用程序，运行效果如图 3-5 所示，这正是我们期望的结果。

图 3-5　修改后的 First 应用程序的运行效果

如前所述，在 Android 应用开发实践中，一般将字符串资源统一定义在 res/values 目录下。因此，我们在 res/values/strings.xml 文件中添加一行代码，具体如下：

```xml
<resources>
    <string name="app_name">First</string>
    <string name="hello">你好，世界！</string>
    <string name="started">Android开发入门了！</string>
    <string name="group">欢迎进入Android阵营！</string>
</resources>
```

同时修改 MainActivity.java 文件中的 onCreate()方法，修改后的代码如下：

```java
public class MainActivity extends AppCompatActivity {

    @Override
    protected void onCreate(Bundle savedInstanceState) {
        super.onCreate(savedInstanceState);
        setContentView(R.layout.activity_main);
        TextView tv = findViewById(R.id.tv02);
        //tv.setText("欢迎进入Android阵营！");
        tv.setText(R.string.group);
```

        }
    }

运行修改后的 First 应用程序，会得到同样的结果，但是这种编码方式更符合 Android 的开发规范。

3.3.4 图片资源

为了保证 Android 应用程序的可用性和美观性，一般会在 Android 应用程序中使用图片装饰界面。

Android 将图片也定义为资源，它将图片资源作为文件存储于 res/mipmap（或 res/drawable）目录下，并且可以针对不同的手机屏幕尺寸使用不同分辨率的图片资源文件。为什么会这样呢？为了弄明白这一点，有一个事实需要读者了解：Android 是一个支持国际化和多设备类型的平台，为了支持国际化和多设备类型，Android 会对代码与资源进行分离管理。那么，如何使一个资源能同时满足多种语言及设备类型呢？Android 通过采用资源目录后缀的方式解决这个问题。例如，为不同的屏幕分辨率和屏幕大小创建不同的图片资源文件，对于高分辨率的设备，会自动从 res/mipmap-hdpi 目录下获取图片资源文件；对于低分辨率的设备，会自动从 res/mipmap-ldpi 目录下获取图片资源文件；对于中等分辨率的设备，会自动从 res/mipmap-mdpi 目录下获取图片资源文件。-xhdpi、xxhdpi、-hdpi、-ldpi、-mdpi 等是资源目录后缀，没有后缀的资源目录 mipmap 是默认资源目录。Android 在搜索资源时，会先获取符合条件的资源，若没有符合条件的资源，则会从默认资源目录下获取资源。

如果读者愿意，那么完全可以删除 mipmap-ldpi、mipmap-mdpi 和 mipmap-hdpi 等目录，然后在 res 目录下新建一个名为 mipmap 的图片资源目录，程序也可以正常运行。我们在不删除 mipmap-ldpi、mipmap-mdpi 和 mipmap-hdpi 等目录的情况下，在 res 目录下，新建一个 mipmap 目录，如图 3-6 所示。

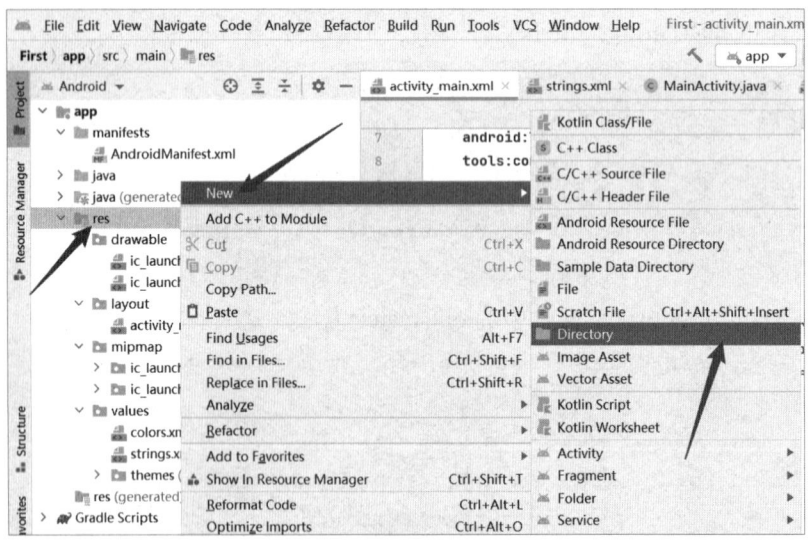

图 3-6　在 res 目录下新建 mipmap 目录

在弹出的对话框中输入"mipmap"，即可创建默认的 mipmap 目录。

图片资源可以是目前 Android 支持的任意图片资源文件，包括.jpg 文件、.png 文件、.bmp

文件等。我们可以将需要使用的图片资源文件直接复制到 res/mipmap 目录下。例如，将 png2030.png 图片资源文件复制到 res/mipmap 目录下，如图 3-7 所示。

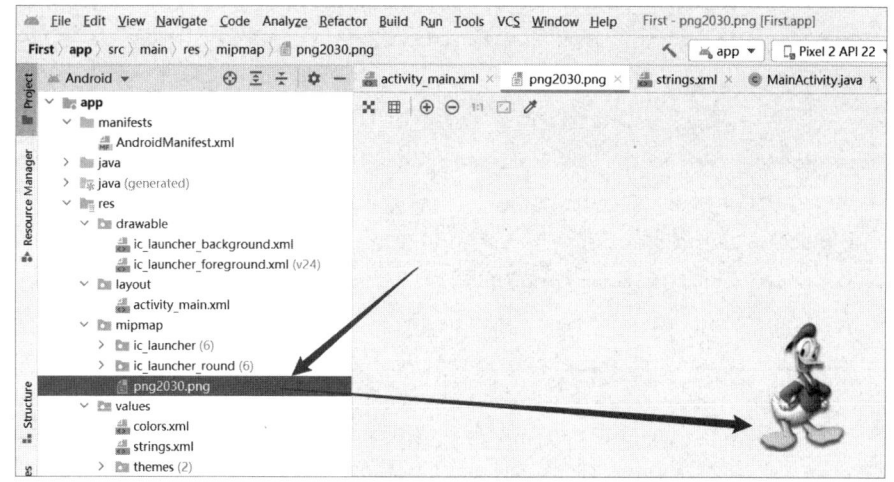

图 3-7　将 png2030.png 图片资源文件复制到 res/mipmap 目录下

如何使用已经添加到 Android 应用程序工程中的图片资源文件呢？最直接的方法是将这个图片资源文件在屏幕上显示出来：我们将它显示在 First 应用程序的 TextView 组件的下方。因此，我们需要对布局文件 res/layout/activity_main.xml 进行修改，修改后的代码如下：

```xml
<?xml version="1.0" encoding="utf-8"?>
<androidx.constraintlayout.widget.ConstraintLayout
    xmlns:android="http://schemas.android.com/apk/res/android"
    xmlns:app="http://schemas.android.com/apk/res-auto"
    xmlns:tools="http://schemas.android.com/tools"
    android:layout_width="match_parent"
    android:layout_height="match_parent"
    tools:context=".MainActivity">


    <TextView
        android:id="@+id/tv01"
        android:layout_width="wrap_content"
        android:layout_height="wrap_content"
        android:text="@string/hello"
        app:layout_constraintTop_toTopOf="parent"
        app:layout_constraintLeft_toLeftOf="parent"
        app:layout_constraintRight_toRightOf="parent"
        />


    <TextView
        android:id="@+id/tv02"
        android:layout_width="wrap_content"
```

```
    android:layout_height="wrap_content"
    android:text="@string/started"
    app:layout_constraintTop_toBottomOf="@id/tv01"
    app:layout_constraintLeft_toLeftOf="parent"
    app:layout_constraintRight_toRightOf="parent" />

<ImageView
    android:layout_width="wrap_content"
    android:layout_height="wrap_content"
    android:src="@mipmap/png2030"
    android:contentDescription="a duck"
    app:layout_constraintTop_toBottomOf="@id/tv02"
    app:layout_constraintLeft_toLeftOf="parent"
    app:layout_constraintRight_toRightOf="parent" />

</androidx.constraintlayout.widget.ConstraintLayout>
```

运行修改后的 Android 应用程序，运行效果如图 3-8 所示。

图 3-8　修改后的 Android 应用程序的运行效果

我们将在后续章节中详细介绍常用组件的使用方法。

3.3.5　Android 中的其他资源

除了前面介绍的资源，Android 中还有多种其他类型的资源，包括 color、animation、array、color-drawable 和 raw 等，我们将在后续章节中对这些资源进行详细介绍。

3.3.6　引用资源

所有的 Android 应用程序资源都是通过与之关联的标识符进行引用的。例如，在 Java 代码中，使用以下代码，可以引用 message_1 字符串资源。

```
R.string.message_1
```

使用以下代码，可以引用 png2030.png 图片资源文件。

```
R.drawable.png2030
```

类似地，在 XML 资源配置文件中，使用以下代码，可以引用 message_1 字符串资源。

```
@string/message_1
```

使用以下代码，可以引用 png2030.png 图片资源文件。

```
@drawable/png2030
```

在一般情况下，在 Java 代码中，引用 Android 应用程序资源的语法格式如下：

```
[package.]R.type.name
```

类似地，在 XML 资源配置文件中，引用 Android 应用程序资源的语法格式如下：

```
@[package:]type/name
```

其中的 type 表示 Android 编译时生成的临时资源类 R 的资源类型，具体如下。
- drawable：图片资源。
- id：在布局中为组件赋予的 id。
- layout：布局资源。
- string：字符串资源。
- menu：菜单资源。
- string-array：字符串数组资源。

name 是各个资源的名称。需要注意的是，资源（如布局资源、图片资源）的名称可能是具体的文件名，也可能是某个资源文件中为具体资源赋予的名称，如字符串资源。

在引用 Android 应用程序资源时，若没有指明 package，则默认引用 Android 应用程序中自定义的资源。Android 平台已经定义了大量资源供用户使用，可以在 Android 帮助文档中看到。打开 Android 开发者网站，进入帮助文档页面，在搜索栏中搜索 "R"，单击搜索结果中的 R.string 超链接，如图 3-9 所示，即可看到 android.R.string 类中定义的字符串资源，如图 3-10 所示。

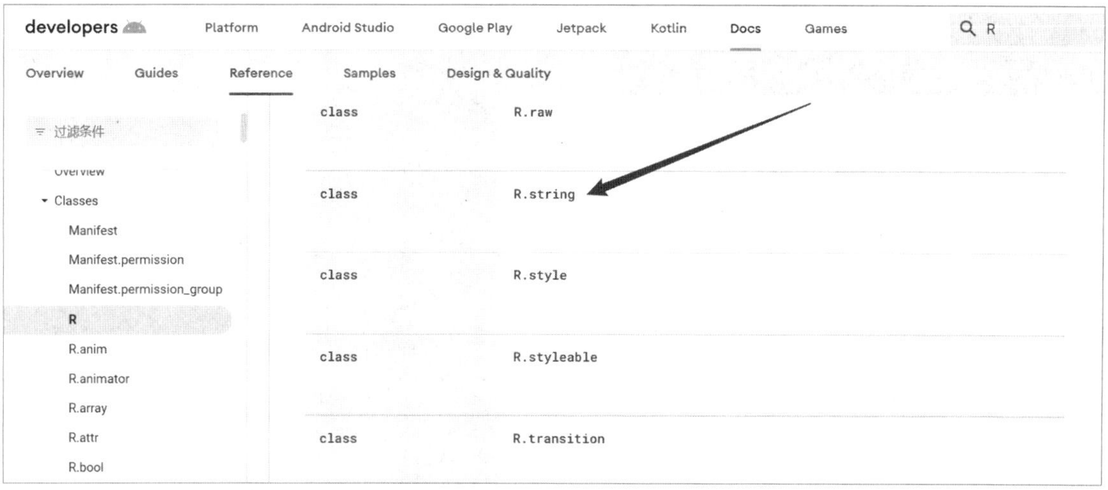

图 3-9　Android 平台自带的资源

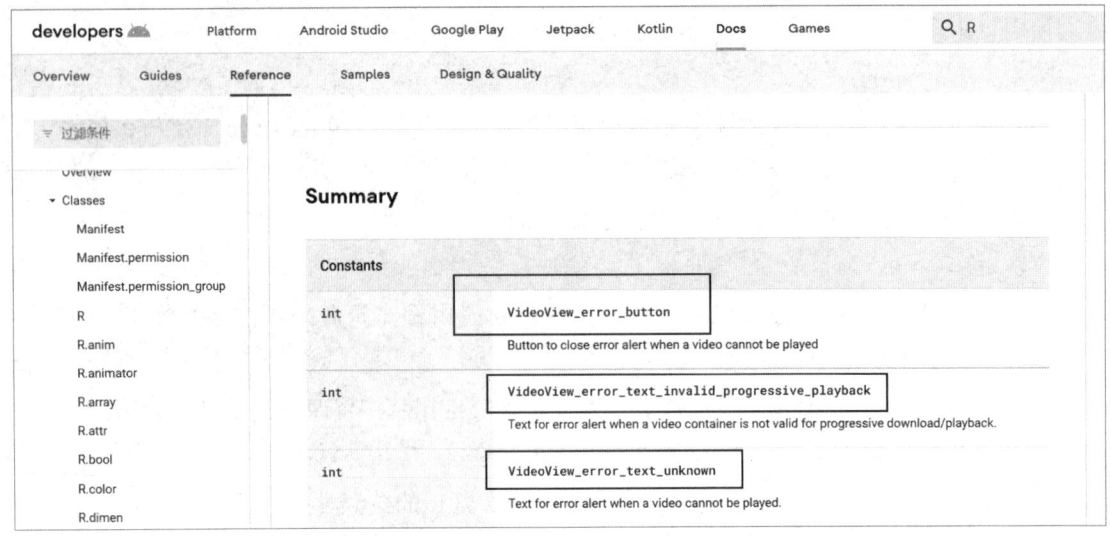

图 3-10　android.R.string 类中定义的字符串资源

在图 3-10 中，VideoView_error_button 是一个字符串常量。在 Java 代码中，可以使用以下代码引用该资源。

```
android.R.string.VideoView_error_button
```

在 XML 资源配置文件中，可以使用以下代码引用该资源。

```
@android:string/VideoView_error_button
```

3.4　同步练习

1. 编写一个 Android 应用程序，并且显示一张自己认为漂亮的图片，然后以文本的形式介绍图片中的内容。

2. 在 Android 帮助文档中找出 Android 平台定义的所有图片资源文件，并且在 Android 应用程序中显示其中任意一张图片。

第 4 章

深入分析 Activity

4.1 Activity 的生命周期

如前所述，Activity 是 Android 的重要组成部分，它是 Android 应用程序的界面。我们在编写 C 语言或 Java SE 应用程序时，都会有一个 main()方法，这个 main()方法就是应用程序的入口。在之前编写的 First 应用程序工程中，MainActivity.java 文件中的代码如下：

```java
package com.example.first;

import androidx.appcompat.app.AppCompatActivity;

import android.os.Bundle;
import android.widget.TextView;

public class MainActivity extends AppCompatActivity {

    @Override
    protected void onCreate(Bundle savedInstanceState) {
        super.onCreate(savedInstanceState);
        setContentView(R.layout.activity_main);
        TextView tv = findViewById(R.id.tv02);
        //tv.setText("欢迎进入Android阵营！");
        tv.setText(R.string.group);
        //tv.setText(android.R.string.VideoView_error_button);
    }
}
```

上述代码中没有 main()方法，First 应用程序是如何运行的呢？这需要我们理解应用程序生命周期的概念。

对于生命周期这个词，我们都不陌生，任何有生命的东西都有生命周期。例如，一棵树是有生命周期的，它的生命周期是种子→发芽→长成小树→长成大树→死亡；一条鱼也是有生命周期的，它的生命周期是鱼卵→小鱼→成年鱼→死亡。同理，一个应用程序也是有生命周期的，它的生命周期是启动应用程序→完成业务功能→应用程序运行结束。需要注意的是，每个事物的生命周期都被某个其他事务管理着：小树的生命周期被大自然管理着，小鱼的生命周期也被大自然管理着，应用程序的生命周期被计算机系统管理着。Android 应用程序的 Activity 也是有生命周期的，它的生命周期是 Android 平台为显示界面做准备→显示界面→用户与界面交互→关闭界面。其实，Android 应用程序的 Activity 的生命周期要比上面描述的复杂和细致得多，并且，Android 应用程序的生命周期是由 Android 平台管理的。Android 应用程序的 Activity 的生命周期的完整过程如图 4-1 所示。

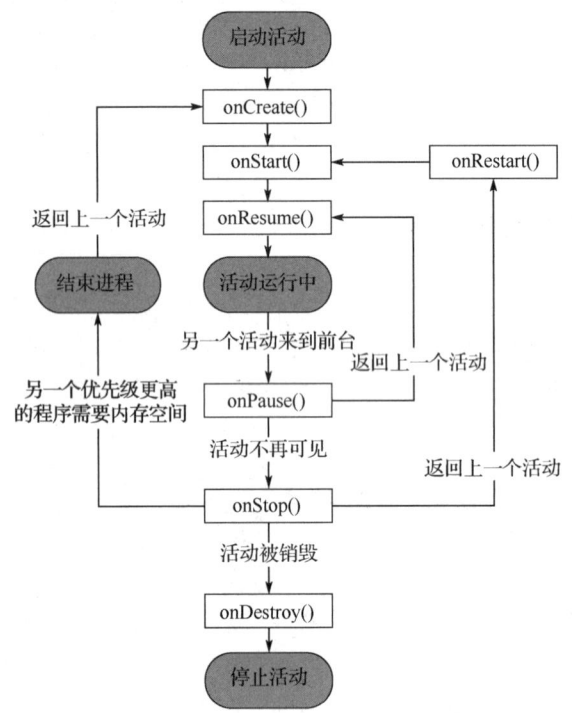

图 4-1 Android 应用程序的 Activity 的生命周期的完整过程

图 4-1 中的 onXXX()称为生命周期方法或生命周期状态，Android 平台会自动调用 Activity 提供的这些生命周期方法，使 Activity 在适当的时候处于适当的状态。这句话的含义有些难理解，没有关系，下面我们对它进行详细的讲解。

根据图 4-1 可知，一个 Activity 从启动到停止要经过多个状态，期间，Android 平台会根据 Activity 状态的变化，调用 Activity 的特定生命周期方法。

如果 Android 平台要启动一个应用程序的 Activity（如启动某个 Android 应用程序的入口 Activity），那么首先调用该 Activity 的 onCreate()方法，然后调用该 Activity 的 onStart()方法，最后调用 Activity 的 onResume()方法，在调用 onResume()方法后，该 Activity（界面）会显示于屏幕上，此时，用户可以与此 Activity 进行交互。例如，对于打电话的应用程序，可以输入要拨打的电话号码。

一个应用程序的 Activity 在显示于屏幕上后，随时都有可能被其他 Activity 覆盖（其他界面）。例如，当进行语音通话时，可能会突然收到一条短信，可以在通话的同时，打开短信应用程序阅读收到的短信，此时，语音通话应用程序的 Activity 被短信应用程序的 Activity 覆盖。在这种情况下，Android 平台会自动调用语音通话应用程序的 Activity 的 onPause()方法，然后调用短信应用程序的入口 Activity 的 onCreate()方法、onStart()方法、onResume()方法，此时，就可以查看短信了；如果语音通话应用程序的 Activity 被完全覆盖，则会调用语音通话应用程序的 Activity 的 onStop()方法。在阅读完短信并关闭短信应用程序后，Android 平台会根据语音通话应用程序的 Activity 被覆盖的情况（部分或完全被短信界面覆盖），自动调用语音通话应用程序的 Activity 的 onResume()方法，或者首先调用 onRestart()方法，然后调用 onStart()方法，最后调用 onResume()方法，将语音通话应用程序的 Activity 显示于屏幕上。

当用户调用 Activity 的 finish()方法时，Android 平台会自动按顺序调用该 Activity 的 onPause()→onStop()→onDestroy()方法，用于关闭这个 Activity，停止运行当前应用程序。例如，当打完电话要关闭电话应用程序界面时，Android 平台会自动按顺序调用电话应用程序的 Activity 的 onPause()→onStop()→onDestroy()方法，用于关闭这个 Activity，停止运行电话应用程序。

4.2 Activity 生命周期案例

修改之前的 MainActivity.java 文件，用于观察 Activity 生命周期的变化，修改后的代码如下：

```java
package com.example.first;

import androidx.appcompat.app.AppCompatActivity;

import android.os.Bundle;
import android.widget.TextView;

public class MainActivity extends AppCompatActivity {

    @Override
    protected void onCreate(Bundle savedInstanceState) {
        super.onCreate(savedInstanceState);
        System.out.println("onCreate() called");

        setContentView(R.layout.activity_main);
        TextView tv = findViewById(R.id.tv02);
        //tv.setText("欢迎进入Android阵营！");
        tv.setText(R.string.group);
        //tv.setText(android.R.string.VideoView_error_button);
    }

    @Override
    protected void onStart() {
        super.onStart();
        System.out.println("onStart() called");
    }

    @Override
```

```
    protected void onResume() {
        super.onResume();
        System.out.println("onResume() called");
    }

    @Override
    protected void onPause() {
        super.onPause();
        System.out.println("onPause() called");
    }

    @Override
    protected void onStop() {
        super.onStop();
        System.out.println("onStop() called");
    }

    @Override
    protected void onRestart() {
        super.onRestart();
        System.out.println("onRestart() called");
    }

    @Override
    protected void onDestroy() {
        super.onDestroy();
        System.out.println("onDestroy() called");
    }
}
```

在修改后的代码中,我们在每个生命周期方法中都加入了一些输出语句,用于将一些信息输出到屏幕上,以便观察 Activity 的哪些生命周期方法被调用了。启动并运行 First 应用程序,运行效果如图 4-2 所示。

图 4-2 修改后的 First 应用程序的运行效果

第 4 章 深入分析 Activity

单击 Android Studio 下方的 Logcat 按钮，如图 4-3 所示，打开 Android Studio 的 Logcat 视图，即可看到 MainActivity.java 程序输出的信息，如图 4-4 所示。

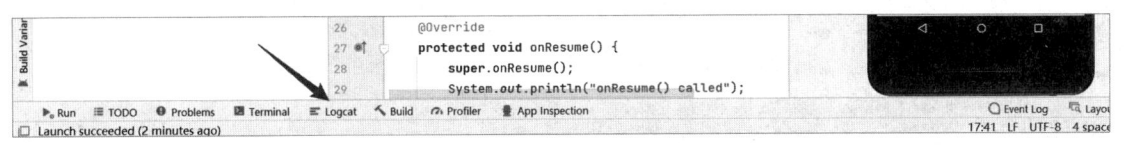

图 4-3　单击 Android Studio 下方的 Logcat 按钮

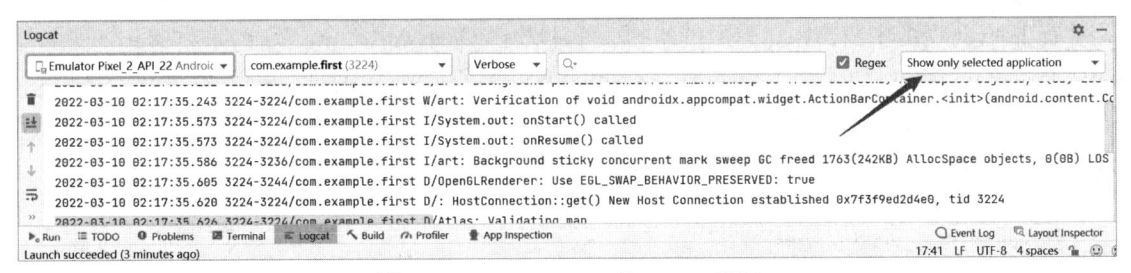

图 4-4　Android Studio 的 Logcat 视图

此时在 Logcat 视图中显示的信息多而杂乱。为了只显示我们需要的信息，在图 4-4 中箭头指示的下拉列表中选择 Edit Filter Configuration 选项，弹出 Create New Logcat Filter 对话框，如图 4-5 所示。

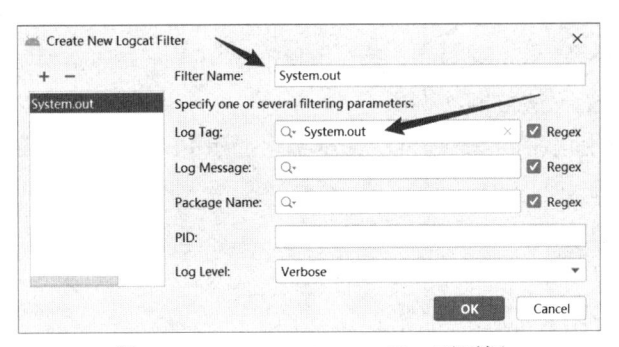

图 4-5　Create New Logcat Filter 对话框

在图 4-5 中填写相关信息后，单击 OK 按钮，即可在 Logcat 视图中只输出我们需要的信息，如图 4-6 所示。

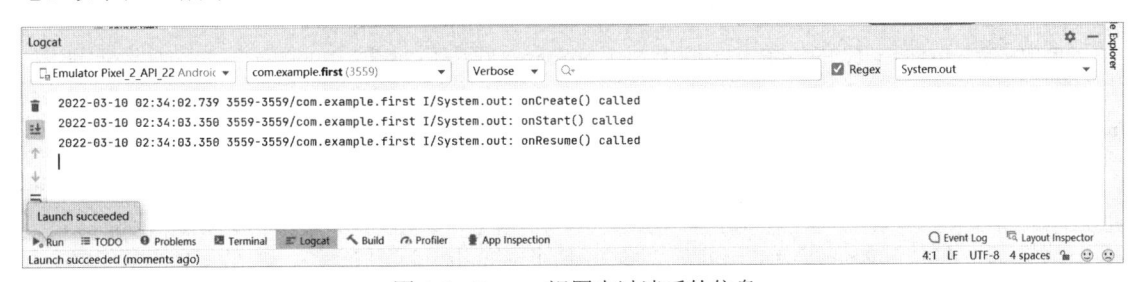

图 4-6　Logcat 视图中过滤后的信息

根据图 4-6 可知，MainActivity 的 onCreate()方法、onStart()方法和 onResume()方法被顺次调用。如果退出 First 应用程序，那么在 Logcat 视图中显示的信息如图 4-7 所示。

根据图 4-7 可知，在退出 First 应用程序时，MainActivity 的 onPause()方法、onStop()方法和 onDestroy()方法被顺次调用。

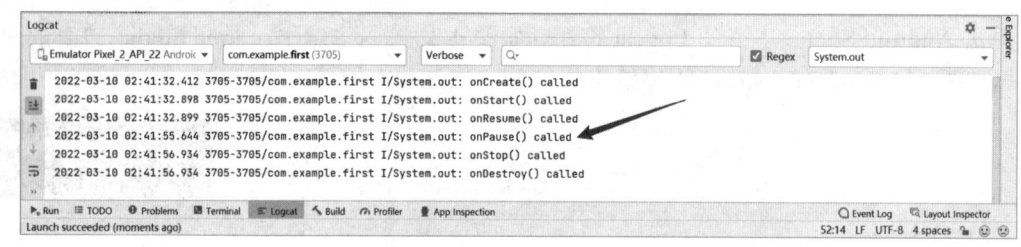

图 4-7　退出 First 应用程序时 Logcat 视图中显示的信息

在 Logcat 视图中右击，在弹出的快捷菜单中选择 Clear All 命令，用于清除 Logcat 视图中的所有信息。再次启动 First 应用程序，在 First 应用程序正常运行后，单击 Android 模拟器下方的圆形按钮以显示 Android 桌面，这时，First 应用程序仍然在运行，只是 MainActivity 被桌面应用程序界面覆盖，这时，Logcat 视图中显示的信息如图 4-8 所示。

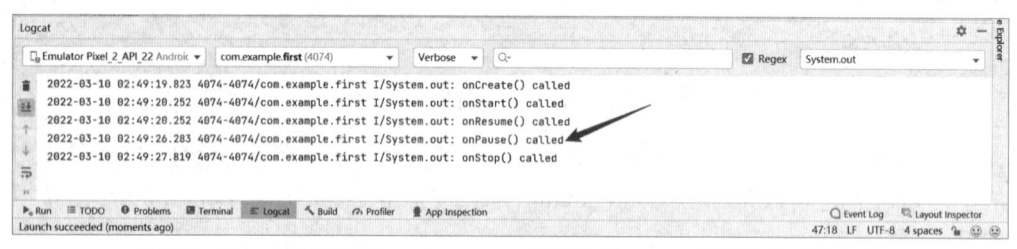

图 4-8　First 应用程序的 MainActivity 被覆盖后 Logcat 视图中显示的信息

由于 First 应用程序的 MainActivity 完全被桌面应用程序界面覆盖，因此，MainActivity 的 onPause()方法、onStop()方法被顺次调用。

再次运行 First 应用程序，也就是再次在屏幕上显示 First 应用程序的 MainActivity，Logcat 视图中显示的信息如图 4-9 所示。

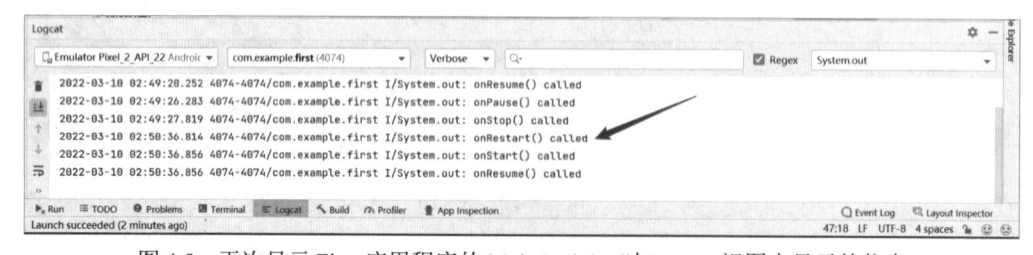

图 4-9　再次显示 First 应用程序的 MainActivity 时 Logcat 视图中显示的信息

此时，MainActivity 的 onRestart()方法、onStart()方法和 onResume()方法再次被顺次调用。

通过这个案例，可以清晰地看到 Activity 的生命周期方法是如何被 Android 平台调用的，理解这个过程是编写完美的 Activity 程序的基础。

在这个案例中，为了方便观察，需要输出一些提示信息。我们使用 Java 自带的 System.out.println()方法输出提示信息。但是，在 Android 应用程序中，一般建议使用 Android 的 Log 类输出信息。下面介绍如何在 Android 应用程序中使用 android.util.Log 类输出程序调试信息。

4.3　使用 Log 类输出程序调试信息

修改 MainActivity.java 文件中的代码，使用 android.util.Log 类输出程序调试信息，修改后的代码如下：

```java
package com.example.first;

import androidx.appcompat.app.AppCompatActivity;

import android.os.Bundle;
import android.util.Log;
import android.widget.TextView;

public class MainActivity extends AppCompatActivity {
    private static String TAG = "我的调试信息";

    @Override
    protected void onCreate(Bundle savedInstanceState) {
        super.onCreate(savedInstanceState);
        Log.i(TAG,"onCreate() called");

        setContentView(R.layout.activity_main);
        TextView tv = findViewById(R.id.tv02);
        //tv.setText("欢迎进入Android阵营！");
        tv.setText(R.string.group);
        //tv.setText(android.R.string.VideoView_error_button);
    }

    @Override
    protected void onStart() {
        super.onStart();
        Log.i(TAG,"onStart() called");
    }

    @Override
    protected void onResume() {
        super.onResume();
        Log.i(TAG,"onResume() called");
    }

    @Override
    protected void onPause() {
        super.onPause();
        Log.i(TAG,"onPause() called");
    }
```

```
    @Override
    protected void onStop() {
        super.onStop();
        Log.i(TAG,"onStop() called");
    }

    @Override
    protected void onRestart() {
        super.onRestart();
        Log.i(TAG,"onRestart() called");
    }

    @Override
    protected void onDestroy() {
        super.onDestroy();
        Log.i(TAG,"onDestroy() called");
    }
}
```

Android 建议为每个 Activity 分配一个 TAG，代码如下：

```
private static String TAG = "我的调试信息";
```

使用 Log 类提供的静态方法显示程序调试信息。例如，使用以下代码，在 Logcat 视图中显示 onCreate()生命周期方法的相关信息。

```
Log.i(TAG, "onCreate");
```

运行修改后的 First 应用程序，并且在 Logcat 视图中创建一个名为"我的调试信息"的信息过滤器，Logcat 视图中输出的程序调试信息如图 4-10 所示。

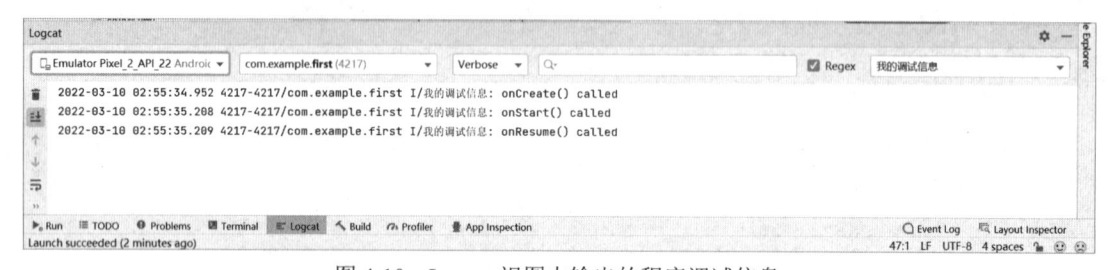

图 4-10 Logcat 视图中输出的程序调试信息

Log 类是一个用于输出程序调试信息的 Android 常用类，它提供了一系列静态方法，常用的方法及其功能如表 4-1 所示。

表 4-1 Log 类中常用的方法及其功能

方法声明	功能
static int v(String tag,String msg)	用 VERBOSE 方式打印消息
static int v(String tag,String msg, Throwable tr)	用 VERBOSE 方式打印消息并输出异常信息
static int d(String tag,String msg)	用 DEBUG 方式打印消息

续表

方法声明	功能
static int d(String tag,String msg, Throwable tr)	用 DEBUG 方式打印消息并输出异常信息
static int i(String tag,String msg)	用 INFO 方式打印消息
static int i(String tag, String msg, Throwable tr)	用 INFO 方式打印消息并输出异常信息
static int w(String tag, Throwable tr)	用 WARN 方式打印异常信息
static int w(String tag, String msg)	用 WARN 方式打印消息
static int w(String tag,String msg, Throwable tr)	用 WARN 方式打印消息并输出异常信息
static int e(String tag, String msg)	用 ERROR 方式打印消息
static int e(String tag,String msg, Throwable tr)	用 ERROR 方式打印消息并输出异常信息
static String getStackTraceString(Throwable tr)	获取异常对象的堆栈信息
static boolean isLoggable(String tag, int level)	指定标签是否可以在指定日志等级中输出
static int println(int priority,String tag, String msg)	将指定的日志消息按指定的日志等级输出
static int wtf(String tag,Throwable tr)	显示错误信息，并且根据系统配置结束当前进程
static int wtf(String tag,String msg)	显示错误信息，并且根据系统配置结束当前进程
static int wtf(String tag, String msg, Throwable tr)	显示错误信息，并且根据系统配置结束当前进程

4.4　Android 中常见的 Activity

Activity 类及其子类是 Android SDK 中的关键类，这些类的继承关系如图 4-11 所示。在后续章节中将介绍这些类的使用方法。

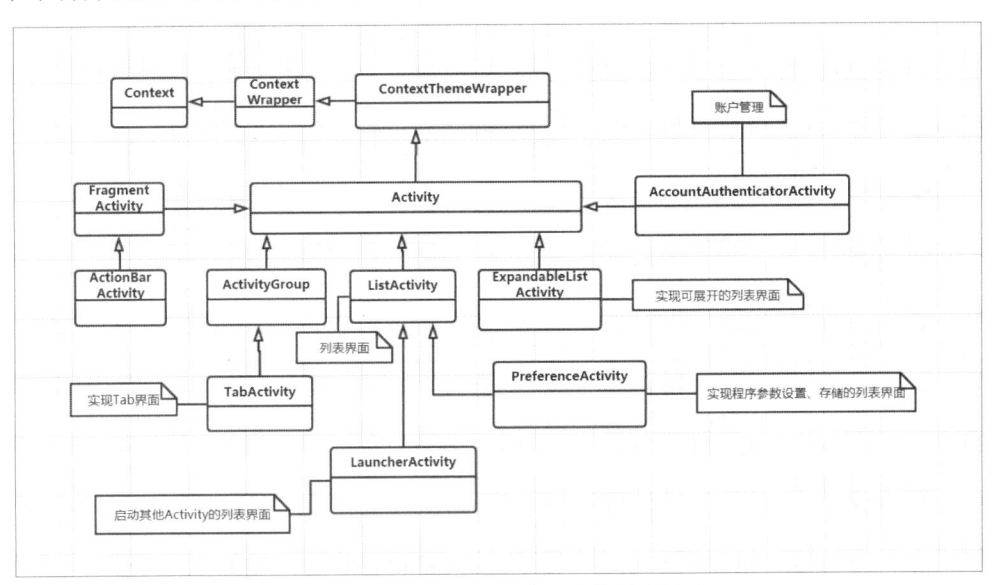

图 4-11　Activity 类及其子类的继承关系

4.5　同步练习

编写一个简单的 Android 应用程序，将 Activity 生命周期方法的调用过程写入一个文件，以便查看 Activity 生命周期的具体过程。

第 5 章
Android 中常用的 UI 组件

5.1 使用基于 XML 的布局

Android 建议采用将应用程序界面与应用程序业务逻辑分离的方式设计应用程序，其实，这不是 Android 的首创。在设计网页应用程序时，也可以采用将要显示的信息与页面布局分离的方式：将要显示的信息存储于 HTML 文件中，而在 CSS 文件中定义信息的显示格式。下面通过一个案例展示 Android 基于 XML 的界面布局。

新建一个 Android 应用程序，在这个应用程序中会显示一个巨大的按钮，用户在点击这个按钮后，会在按钮上显示当前的日期和时间。为此，在 Android Studio 中新建一个名为 Ch0501 的 Android 应用程序工程，其结构如图 5-1 所示。

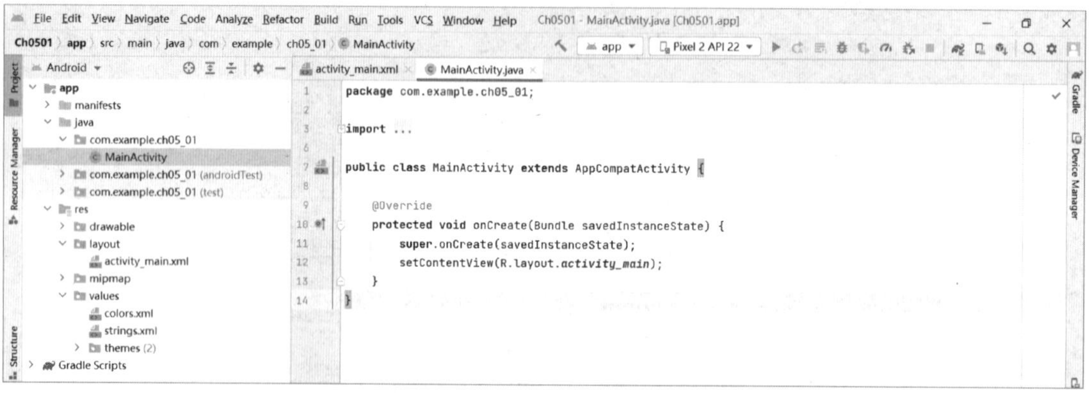

图 5-1　Ch0501 应用程序工程的结构

打开布局文件 activity_main.xml，修改其中的内容，修改后的代码如下：

```xml
<?xml version="1.0" encoding="utf-8"?>
<LinearLayout xmlns:android="http://schemas.android.com/apk/res/android"
    android:layout_width="match_parent"
    android:layout_height="match_parent">

    <android.widget.Button
        android:id="@+id/id_button_01"
        android:layout_width="match_parent"
        android:layout_height="match_parent"
        android:text="@string/text_button_01" />
```

```
</LinearLayout>
```

根据上述代码可知，LinearLayout 布局容器中包含一个按钮，这个按钮的 id 为 id_button_01，该按钮会占据 LinearLayout 布局容器的全部可用显示区域，并且会在按钮上显示由 @string/ text_button_01 引用的字符串资源。由于我们引用了标识符为 text_button_01 的字符串资源，因此需要修改 res/values/strings.xml 文件的内容，在其中定义一个 name 属性值为 text_button_01 的字符串资源，修改后的代码如下：

```
<resources>
    <string name="app_name">Ch0501</string>
    <string name="text_button_01">点击</string>
</resources>
```

运行 Ch0501 应用程序，运行效果如图 5-2 所示。

在 Ch0501 应用程序中，按钮可以被点击，在点击后，按钮的颜色会发生变化。本案例要实现的功能为，在点击按钮后，在按钮上显示当前的日期和时间。也就是说，Ch0501 应用程序需要响应用户对按钮的点击操作。那么如何响应用户对按钮的点击操作呢？通过实现 View.OnClickListener 接口达到这个目的。

View.OnClickListener 接口是 android.view.View 类的一个内部接口，如图 5-3 所示。在 View.OnClickListener 接口中定义了组件被点击时的响应方法——onClick()，该方法可以在 Android 帮助文档中看到。如果要快速查看某个 Android 类的相关信息，那么在 Android 帮助文档首页的搜索栏中直接输入要查看的 Android 类的名称，即可显示其相关信息。例如，为了快速查看 OnClickListener 接口的相关信息，可以在 Android 帮助文档首页的搜索栏中输入"OnClickListener"，即可显示该接口的相关信息。

图 5-2 Ch0501 应用程序的运行效果

图 5-3 View.OnClickListener 接口

修改 MainActivity.java 文件中的代码：监听对 id 为 id_button_01 的按钮的点击事件，当该按钮被点击时，对这个点击事件进行相应的响应，修改后的代码如下：

```java
package com.example.ch0501;

import androidx.appcompat.app.AppCompatActivity;

import android.os.Bundle;
import android.view.View;
import android.widget.Button;

import java.util.Date;

public class MainActivity extends AppCompatActivity implements View.OnClickListener{
    Button btn;

    @Override
    protected void onCreate(Bundle savedInstanceState) {
        super.onCreate(savedInstanceState);
        setContentView(R.layout.activity_main);
        //获取界面中Button组件（按钮）的引用
        btn = this.findViewById(R.id.id_button_01);
        btn.setOnClickListener(this);              //监听对该按钮的点击事件

    }

    public void onClick(View v) {
        Date d = new Date();
        btn.setText(d.toString());
    }
}
```

MainActivity 类实现了 View.OnClickListener 接口，在 onCreate()方法的实现中获取界面中 Button 组件（按钮）的引用，并且监听对该按钮的点击事件，代码如下：

```java
//获取界面中Button组件（按钮）的引用
btn = this.findViewById(R.id.id_button_01);
btn.setOnClickListener(this);                      //监听对该按钮的点击事件
```

当该按钮被点击时，会调用 MainActivity 类的 onClick()方法（因为 MainActivity 类实现了 OnClickListener 接口）。在 onClick()方法的实现中，获取当前的日期和时间，并且将其转换为字符串显示在按钮上，代码如下：

```java
    public void onClick(View v) {
        Date d = new Date();
        btn.setText(d.toString());
    }
```

运行修改后的 Ch0501 应用程序，点击按钮，运行效果如图 5-4 所示。

图 5-4　点击按钮后 Ch0501 应用程序的运行效果

不停地点击按钮，每次点击按钮，都会显示当前的日期和时间。

虽然我们没有对 Android 中的组件进行详细的讲解，但是，通过这个案例可以看出，Android 平台是如何将应用程序界面与应用程序业务逻辑分离的。下面详细讲解 Android 中基本组件的相关知识。

5.2　Android 中的基本组件

Android 中的基本组件包括 Button、TextView、ImageView、EditText、CheckBox、RadioButton 等。Android 中的组件位于 android.widget 包中。

5.2.1　Button 组件

Button 组件是常用的组件，读者可以打开 Android 帮助文档，观察 Button 组件的 XML 配置属性。Button 组件的每个 XML 配置属性都有一个与之对应的 setXXX()方法，这说明，既可以使用 XML 配置属性配置组件的属性值，又可以在代码中使用 setXXX()方法设置同样的属性值。

Button 组件没有自己的 XML 配置属性，它的 XML 配置属性都是从父类 TextView 中继承来的（将在下一节介绍 TextView 组件）。Button 组件的常用布局属性如下。

- android:text。设置显示在 Button 组件上的文字，可以是一个字符串常量，也可以是一个字符串资源的引用。按照 Android 的规则，建议不要直接使用字符串常量，应该使用字符串资源的引用。

- android:textColor。设置显示在 Button 组件上的文本颜色，其值可以是一个颜色资源的引用，也可以是一个颜色值，如"#rgb"、"#argb"、"#rrggbb"和"#aarrggbb"。建议使用颜色资源的引用，至于如何定义及引用颜色资源，将在后续章节中进行介绍。
- android:textSize。设置显示在 Button 组件上的文本字号，其值可以是一个常数加单位，如 15px、20sp 等，也可以是一个单位度量资源的引用。建议使用单位度量资源的引用，至于如何定义及引用单位度量资源，将在后续章节中进行介绍。
- android:textStyle。设置显示在 Button 组件上的文本风格，其值包括 bold、italic、bolditalic。
- android:typeface。设置显示在 Button 组件上的文本字体。目前 Android 支持的字体包括 normal、sans、serif、monospace。

以上属性都是从 android.widget.TextView 中继承来的，还有一些 XML 配置属性是从 android.view.View 中继承来的，下面介绍这些属性。

- android:id。设置 Button 组件的 id 属性，以便在程序代码中引用该组件。
- android:background。设置 Button 组件的背景，其值可以是一个 drawable 资源的引用，也可以是一个颜色值，如"#rgb"、"#argb"、"#rrggbb"和"#aarrggbb"。
- android:clickable。设置 Button 组件是否可以响应点击事件，其值包括 true、false。
- android:visibility。设置 Button 组件是否可以显示在屏幕上，其值包括 true、false。
- android:padding、android:paddingTop、android:paddingBottom、android:paddingLeft、android:paddingRight。设置 Button 组件的内边界，类似于 HTML/CSS 中的 padding 属性。
- android:gravity。设置显示在 Button 组件上的提示文字的对齐方式，其值包括 top、right、left、center 等。

将 Ch0501 应用程序工程重命名为 Ch0502，并且对其进行修改，我们希望在按钮上显示一张背景图片。因此，我们需要在布局文件 res/layout/activity_main.xml 中为 Button 组件添加一个 android:background 属性，修改后的 activity_main.xml 文件中的代码如下：

```xml
<?xml version="1.0" encoding="utf-8"?>
<LinearLayout xmlns:android="http://schemas.android.com/apk/res/android"
    android:layout_width="match_parent"
    android:layout_height="match_parent">

    <android.widget.Button
        android:id="@+id/id_button_01"
        android:layout_width="match_parent"
        android:layout_height="match_parent"
        android:background="@mipmap/png0030"
        android:text="@string/text_button_01" />

</LinearLayout>
```

在 Button 组件的 XML 配置属性中指定了按钮的背景图片，即 android:background 配置属性所指定的值：它指定采用一个名为 png0030.png 的图片资源文件作为该按钮的背景图片。回顾之前介绍的知识，需要将名为 png0030.png 的图片资源文件存储于 res/mipmap-xxx 目录下，Ch0502 应用程序工程中包含多个 res/mipmap-xxx 目录，那么，应该将图片资源文件存储于哪个目录下呢？

为了回答这个问题，我们需要知道一个事实：Android 是一个支持多种手持设备的操作系统平台，它既可以支持不同的手机，又可以支持平板电脑及可穿戴设备，这里只讨论手机。为了使同一个应用程序能够在多种不同的手机上使用，需要为不同的手机（如不同的屏幕分辨率、不同的屏幕大小、使用不同的 Android 版本、手机是竖屏还是横屏、不同的语言等）指定不同的资源，Android 通过资源配置量词达到这个目的。res 资源目录下的所有子资源目录（如 mipmap、drawable、layout、menu、values 等，后缀名包括 -ldpi、-mdpi、-hdpi、-xhdpi、-xxhdpi 等）都是资源配置量词。Android 的资源配置量词不止这些，我们将在后续章节中专门介绍常用的资源配置量词。

现在以在按钮上显示背景图片为例，说明如何为不同的手机指定不同的资源，也就是应该将 png0030.png 图片资源文件存储于哪个 mipmap-xxx 目录下。正确的做法是，为具有不同分辨率的手机制作不同尺寸的图片资源文件，给它们赋予同一个文件名，并且分别存储于相应的 mipmap-mdpi、mipmap-hdpi、mipmap-xhdpi、mipmap-xxhdpi 目录下。例如，制作尺寸为 64px×64px 的图片资源文件并将其存储于 mipmap-mdpi 目录下，制作尺寸为 96px×96px 的图片资源文件并将其存储于 mipmap-hdpi 目录下，制作尺寸为 128px×128px 的图片资源文件并将其存储于 mipmap-xhdpi 目录下，制作尺寸为 256px×256px 的图片资源文件并将其存储于 mipmap-xxhdpi 目录下，这样，当应用程序在具有不同分辨率的手机上运行时，所显示的图片大小看起来都差不多。

将制作完成的不同尺寸的 png0030.png 图片资源文件分别存储于不同的 mipmap-xxx 目录下，运行 Ch0502 应用程序，运行效果如图 5-5 所示。

在继续介绍 Android 中的基本组件前，先介绍一下 Android 的常用资源配置量词，如表 5-1 所示。

图 5-5　Ch0502 应用程序的运行效果

表 5-1　Android 的常用资源配置量词

类别	资源配置量词	含义
屏幕分辨率	ldpi	低密度屏幕分辨率，120dpi（120 点/英寸）
	mdpi	中密度屏幕分辨率，160dpi（160 点/英寸）
	hdpi	高密度屏幕分辨率，240dpi（240 点/英寸）
	xhdpi	超高密度屏幕分辨率，320dpi（320 点/英寸）
	xxhdpi	超超高密度屏幕分辨率，480dpi（480 点/英寸）
	xxxhdpi	超超超高密度屏幕分辨率，640dpi（640 点/英寸）
屏幕大小	small	屏幕大小至少是 320dp×426dp。需要注意的是，这里的 dp 不是屏幕的物理像素，而是逻辑像素。dp 的英文含义是 density independent pixel，在分辨率为 160dpi 的屏幕上，1dp = 1px；在分辨率为 240dpi 的屏幕上，1dp=1.5px；以此类推
	normal	屏幕大小至少是 320dp×470dp
	large	屏幕大小至少是 480dp×640dp
	xlarge	屏幕大小至少是 720dp×960dp

续表

类别	资源配置量词	含义
屏幕方位	port	手机处于竖屏状态，即手机是被竖着（Portrait）拿的
	land	手机处于横屏状态，即手机是被横着（Landscape）拿的
API 版本	v4	API 版本号为 4
	v7	API 版本号为 7
	v11	API 版本号为 11
	v14	API 版本号为 14
	等	—
语言和地区	cn	中文
	en	英文
	fr	法文
	fr-rCA	法文，加拿大地区
	等	—

此时再回头看前面编写的工程结构，会发现有多个 values 资源目录，分别是 values、values-v11、values-v14，也就是说，在 API 版本号为 11 的 Android 手机上，使用 values-v11 资源目录下的资源；在 API 版本号大于或等于 14 的 Android 手机上，使用 values-v14 资源目录下的资源；在其他 API 版本号的 Android 手机上，默认使用 values 资源目录下的资源。

5.2.2 TextView 组件

TextView 组件类似于 GUI 中的 Label 组件，主要用于显示一个字符串，显示在 TextView 组件中的字符串是不可编辑的。其实，Button 组件的 XML 配置属性都是从 TextView 组件和 View 组件中继承的，因此，TextView 组件的 XML 配置属性与 Button 组件的 XML 配置属性是类似的。

修改 Ch0502 应用程序工程中的代码，将 Button 组件替换为 TextView 组件，修改后的 res/layout/activity_main.xml 文件中的代码如下：

```xml
<?xml version="1.0" encoding="utf-8"?>
<LinearLayout xmlns:android="http://schemas.android.com/apk/res/android"
    android:layout_width="match_parent"
    android:layout_height="match_parent">

    <android.widget.TextView
        android:id="@+id/textview"
        android:layout_width="match_parent"
        android:layout_height="match_parent"
        android:text="@string/hello_world"
        android:textColor="#A08800FF"
        android:textSize="15sp"
        android:textStyle="bold"
        android:typeface="monospace"
        android:background="@mipmap/png0030" />

</LinearLayout>
```

修改后的 MainActivity.java 文件中的代码如下：

```java
package com.example.ch0502;

import androidx.appcompat.app.AppCompatActivity;

import android.os.Bundle;
import android.widget.Button;

public class MainActivity extends AppCompatActivity{
    Button btn;

    @Override
    protected void onCreate(Bundle savedInstanceState) {
        super.onCreate(savedInstanceState);
        setContentView(R.layout.activity_main);
    }
}
```

修改后的 res/values/strings.xml 文件中的代码如下：

```xml
<resources>
    <string name="app_name">Ch0502</string>
    <string name="text_button_01">点击</string>
    <string name="hello_world">Hello World</string>
</resources>
```

运行修改后的 Ch0502 应用程序，运行效果如图 5-6 所示。

图 5-6　修改后的 Ch0502 应用程序的运行效果

5.2.3 ImageView 组件

ImageView 组件主要用于显示一张图片。ImageView 组件常用的 XML 配置属性如下。
- android:maxHeight。指定组件的最大高度。
- android:maxWidth。指定组件的最大宽度。
- android:scaleType。控制显示在 ImageView 组件中的图片应该如何改变大小，从而适应 ImageView 组件的大小，可用的值及其含义如下。
  - center：图片位于视图中间，但不进行缩放。
  - centerCrop：按统一比例缩放图片（保持图片的尺寸比例），使图片的两维（宽度和高度）大于或等于相应视图的维度。
  - centerInside：按统一比例缩放图片（保持图片的尺寸比例），使图片的两维（宽度和高度）大于或等于相应视图的维度。
  - fitCenter：按比例缩放图片，使其大小达到组件的大小，并且使图片居中显示。
  - fitEnd：按比例缩放图片，使其大小达到组件的大小，并且使图片居末显示。
  - fitStart：按比例缩放图片，使其大小达到组件的大小，并且使图片居首显示。
  - fitXY：不按比例缩放图片，使其两维正好达到组件的维数。
  - matrix：当绘制时，使用图片矩阵变换对图片进行缩放。
- android:src。指定显示在 ImageView 组件中的图片，必须是一个图片资源文件的引用。
- android:contentDescription。设置图片的描述性文字。

5.2.4 EditText 组件

EditText 组件是可编辑的文本组件，与 TextView 组件类似，但是提供了编辑功能。EditText 组件的 XML 配置属性都是从 TextView 组件及 View 组件中继承的，常用的 XML 配置属性如下。
- android:autoText。设置是否对输入的文字进行自动拼写检查，只能取值 true 或 false。
- android:captalize。设置是否将输入的文字改为大写，只能取值 true 或 false。
- android:digits。设置是否只能输入数字，只能取值 true 或 false。
- android:singleLine。设置是否可以输入多行，只能取值 true 或 false。
- android:hint。设置当输入框为空时，在输入框中显示的提示信息。
- android:inputType。设置 EditText 组件中的文字类型，其值包括 none、text、textCapCharacters、textCapWords、textUri、number 等。

5.2.5 CheckBox 组件

CheckBox 组件就是 GUI 中的复选框，它继承了 TextView 组件和 View 组件的 XML 配置属性，该组件的常用方法如下。
- isChecked()。检查当前复选框是否被勾选。
- setChecked(Boolean checked)。设置当前复选框的勾选状态。
- toggle()。将该复选框的勾选状态反选。

5.2.6 RadioButton 组件

RadioButton 组件就是 GUI 中的单选按钮。在一般情况下，将 RadioButton 组件与 RadioGroup

组件结合使用,通过 RadioGroup 组件控制 RadioButton 组件的选中状态,使一组 RadioButton 组件中只可以选中一个。RadioGroup 组件的常用方法如下。
- check(int rb)。检查指定单选按钮的选中状态。
- clearCheck()。清除所有单选按钮的选中状态。因此,在调用该方法后,没有单选按钮被选中。
- getCheckedRadioButtonId()。返回被选中的单选按钮的 id,如果没有单选按钮被选中,则返回-1。

5.3 同步练习一

编写一个简单的 Android 应用程序,在主界面中显示一个按钮,点击该按钮,会在该按钮上以符合中国人阅读习惯的方式显示日期和时间,如以 2022 年 03 月 10 日 10:30:17 的格式显示日期和时间。

5.4 Android 中的容器组件

Android 中的容器组件是指可以放置其他组件,并且可以对放置在其中的组件进行布局的 Android 组件。Android 中常用的容器组件包括 LinearLayout、RelativeLayout、FrameLayout、ScrollView 及 ConstraintLayout。

5.4.1 LinearLayout 容器组件

LinearLayout 容器组件是线性布局组件,放置在其中的组件会按列或按行进行顺序布局。下面介绍 LinearLayout 容器组件常用的 XML 配置属性。
- android:orientation。设置 LinearLayout 容器组件的布局方式:按行或按列,其值为 horizontal 和 vertical。
- android:gravity。设置布局在 LinearLayout 容器组件内组件的对齐方式,其值包括 top、bottom、left、right、center、start、end 等。
- 其他从 View 组件中继承的属性,包括 android:background、android:visibility 等。

对于放置在 LinearLayout 容器组件中的组件,LinearLayout 容器组件提供了一些 XML 配置属性,用于告知 LinearLayout 容器组件如何放置这些组件。这些 XML 配置属性位于 LinearLayout.LayoutParams 配置属性列表中,具体如下。
- android:layout_width 和 android:layout_height。所有放置在 LinearLayout 容器组件中的组件都必须通过 android:layout_width 和 android:layout_height 属性告知 LinearLayout 容器组件如何对其内部的组件进行布局。这两个属性都有以下 3 个可选的值。
  - match_parent/fill_parent。表示占满父容器组件中的所有空间。
  - wrap_content。表示只占用显示器内容所需的空间。
  - 一个常数值和单位。如 100px,表示该组件占用 100 个物理像素。常数值可用的单位如下。
    - px(物理像素):屏幕上的物理点。

- in（英寸）：长度单位。
- mm（毫米）：长度单位。
- pt（磅）：1/72 英寸。
- dp（逻辑像素）：一种基于屏幕密度的抽象单位。在分辨率为 160dpi 的显示器上，1dp = 1px；在分辨率为 240dpi 的显示器上，1dp = 1.5px。
- dip：与 dp 相同，通常用于 Google 案例中。
- sp（与刻度无关的像素）：与 dp 类似，但是可以根据用户的文本字号首选项进行缩放。

- android:layout_gravity。设置组件在 LinearLayout 容器组件中的布局方式。
- android:layout_weight。设置组件占用 LinearLayout 容器组件中空余显示空间的比例。
- android:layout_margin、android:layout_marginTop、android:layout_marginBottom、android:layout_marginLeft、android:layout_marginRight。设置组件的外边界，类似于 HTML/CSS 中的 margin 属性。

下面举例说明 LinearLayout 容器组件的使用方法。在 Android Studio 中新建一个名为 Ch0503 的 Android 应用程序工程，修改布局文件 activity_main.xml 中的代码，修改后的代码如下：

```xml
<?xml version="1.0" encoding="utf-8"?>
<LinearLayout xmlns:android="http://schemas.android.com/apk/res/android"
    android:layout_width="match_parent"
    android:layout_height="match_parent"
    android:orientation="vertical">

    <RadioGroup
        android:id="@+id/orientation"
        android:layout_width="wrap_content"
        android:layout_height="wrap_content"
        android:orientation="horizontal"
        android:padding="5dip" >

        <RadioButton
            android:layout_width="wrap_content"
            android:layout_height="wrap_content"
            android:id="@+id/horizontal"
            android:text="@string/horizontal" />

        <RadioButton
            android:layout_width="wrap_content"
            android:layout_height="wrap_content"
            android:id="@+id/vertical"
            android:text="@string/vertical" />
    </RadioGroup>
```

```xml
<RadioGroup
    android:id="@+id/gravity"
    android:layout_width="match_parent"
    android:layout_height="wrap_content"
    android:orientation="vertical"
    android:padding="5dip" >

    <RadioButton
        android:layout_width="wrap_content"
        android:layout_height="wrap_content"
        android:id="@+id/left"
        android:text="@string/left" />

    <RadioButton
        android:layout_width="wrap_content"
        android:layout_height="wrap_content"
        android:id="@+id/center"
        android:text="@string/center" />

    <RadioButton
        android:layout_width="wrap_content"
        android:layout_height="wrap_content"
        android:id="@+id/right"
        android:text="@string/right" />
</RadioGroup>

</LinearLayout>
```

在 activity_main.xml 文件中，在 LinearLayout 容器组件中嵌套了两个 RadioGroup 组件，这两个 RadioGroup 组件分别用于管理各自的 RadioButton 组件，保证在任何时候都只有一个 RadioButton 组件被选中。需要注意的是，RadioGroup 组件也是一个容器组件，它是 LinearLayout 容器组件的子组件，因此可以在 RadioGroup 组件中放置 RadioButton 组件。为了能够在程序代码中获得各个组件的引用，我们给每个组件都赋予了一个 id。

修改 res/values/strings.xml 文件中的代码，修改后的代码如下：

```xml
<?xml version="1.0" encoding="utf-8"?>
<resources>
    <string name="app_name">Ex05LinearLayout</string>

    <string name="horizontal">水平</string>
    <string name="vertical">垂直</string>
    <string name="left">居左</string>
    <string name="center">居中</string>
```

```xml
        <string name="right">居右</string>

</resources>
```

在 strings.xml 文件中定义了布局文件中引用的字符串资源。

修改 **MainActivity.java** 文件中的代码，修改后的代码如下：

```java
package com.example.ch0503;

import androidx.appcompat.app.AppCompatActivity;

import android.os.Bundle;
import android.view.Gravity;
import android.widget.LinearLayout;
import android.widget.RadioGroup;

public class MainActivity extends AppCompatActivity implements RadioGroup.
    OnCheckedChangeListener {
    RadioGroup orientation;
    RadioGroup gravity;

    @Override
    protected void onCreate(Bundle savedInstanceState) {
        super.onCreate(savedInstanceState);
        setContentView(R.layout.activity_main);

        orientation = findViewById(R.id.orientation);
        orientation.setOnCheckedChangeListener(this);
        gravity = findViewById(R.id.gravity);
        gravity.setOnCheckedChangeListener(this);
    }

    public void onCheckedChanged(RadioGroup group, int checkedId) {
        if (checkedId==R.id.horizontal)   //从Android 8.0开始,不建议使用switch case
            orientation.setOrientation(LinearLayout.HORIZONTAL);
        else if (checkedId == R.id.vertical)
            orientation.setOrientation(LinearLayout.VERTICAL);
        else if (checkedId == R.id.left)
            gravity.setGravity(Gravity.START);
        else if (checkedId == R.id.center)
            gravity.setGravity(Gravity.CENTER_HORIZONTAL);
        else if (checkedId == R.id.right)
            gravity.setGravity(Gravity.END);
    }
}
```

在 MainActivity 类的 onCreate()方法中，我们获得了两个 RadioGroup 组件的引用，并且分别设置了对其中的 RadioButton 组件选中状态发生变化时的监听，然后在 OnCheckedChangeListener 接口的 onCheckedChanged()方法中对 RadioButton 组件选中状态的变化进行处理，例如，如果"水平"单选按钮被选中，则调用 RadioGroup 组件的 setOrientation()方法，设置当前 RadioGroup 组件采用水平布局方式；如果"垂直"单选按钮被选中，则调用 RadioGroup 组件的 setGravity()方法，设置当前 RadioGroup 组件采用垂直布局方式。运行 Ch0503 应用程序，运行效果如图 5-7 所示。

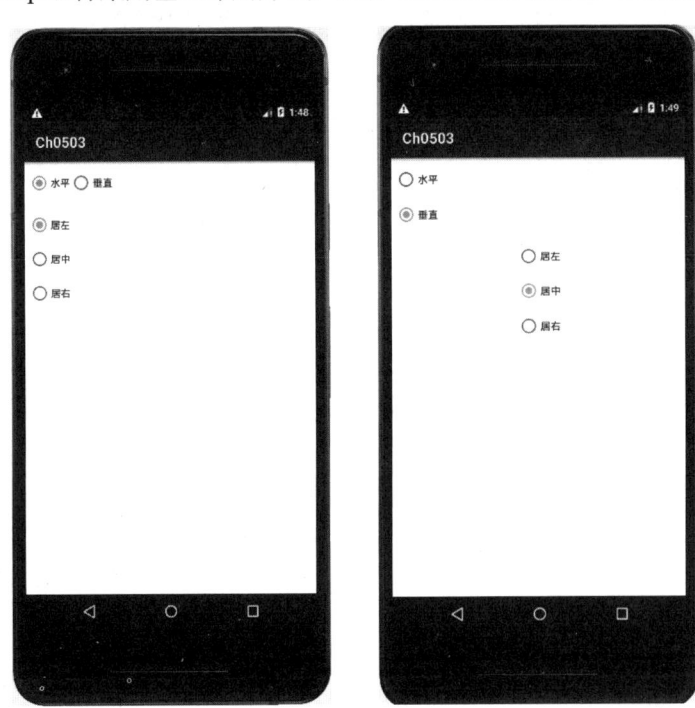

图 5-7　Ch0503 应用程序的运行效果

LinearLayout 容器组件为布局于其中的组件提供了 android:layout_weight 布局属性，它可以使各个组件按指定的比例共享 LinearLayout 容器组件的显示空间。下面举例说明 android:layout_weight 布局属性的使用方法。

在 Android Studio 中新建一个名为 Ch0504 的 Android 应用程序工程，修改布局文件 res/layout/activity_main.xml 中的代码，修改后的代码如下：

```xml
<?xml version="1.0" encoding="utf-8"?>
<LinearLayout xmlns:android="http://schemas.android.com/apk/res/android"
    android:layout_width="match_parent"
    android:layout_height="match_parent"
    android:orientation="vertical">

    <android.widget.Button
        android:layout_width="match_parent"
        android:layout_height="0dp"
        android:layout_weight="5"
        android:text="@string/fifty" />
```

```xml
<android.widget.Button
    android:layout_width="match_parent"
    android:layout_height="0dp"
    android:layout_weight="3"
    android:text="@string/thirty" />

<android.widget.Button
    android:layout_width="match_parent"
    android:layout_height="0dp"
    android:layout_weight="2"
    android:text="@string/twenty" />

</LinearLayout>
```

在 activity_main.xml 文件中，LinearLayout 容器组件中包含 3 个 Button 组件。在采用垂直布局方式的情况下，如果要使用 android:layout_weight 属性，则需要设置 android:layout_height 属性的值为 0dp；在采用水平布局方式的情况下，如果要使用 android:layout_weight 属性，则需要设置 android:layout_width 属性的值为 0dp。在设置完成后，在相应组件的 android:layout_weight 属性中指定这个组件要占用 LinearLayout 容器组件显示空间的比例。在本案例中，将 LinearLayout 容器组件的显示空间平均分为 10 份，其中，第 1 个按钮占 5 份，第 2 个按钮占 3 份，第 3 个按钮占 2 份。

修改 res/values/strings.xml 文件中的代码，在该文件中定义布局文件中引用的字符串资源，代码如下：

```xml
<?xml version="1.0" encoding="utf-8"?>
<resources>
    <string name="app_name">Ch0504</string>

    <string name="fifty">50%</string>
    <string name="thirty">30%</string>
    <string name="twenty">20%</string>
</resources>
```

运行 Ch0504 应用程序，运行效果如图 5-8 所示。

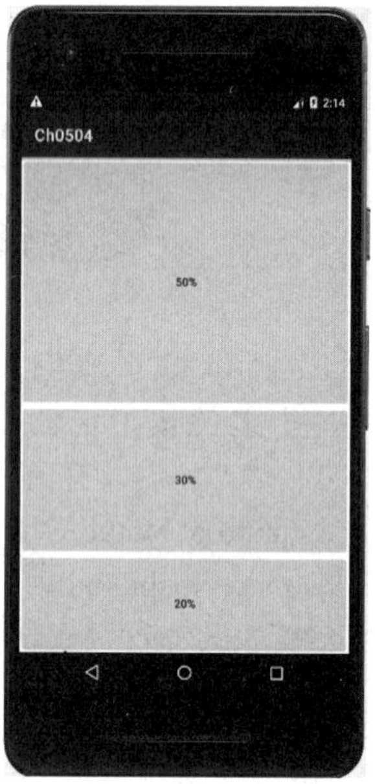

5.4.2 RelativeLayout 容器组件

图 5-8 Ch0504 应用程序的运行效果

顾名思义，RelativeLayout 容器组件是一种采用相对布局方式的容器组件，也就是说，一个组件相对于另一个组件的位置进行布局，如将组件 A 定位到组件 B 的右下方。RelativeLayout 容器组件为布局在其中的组件提供了非常多的 XML 布局属性，常用的 XML 布局属性如表 5-2 所示。

表 5-2　RelativeLayout 容器组件常用的 XML 布局属性

XML 布局属性	含　义
android:layout_above	将组件定位到指定组件的上方
android:layout_alignParentBottom	如果值为 true，则将组件定位到父容器组件的最下方
android:layout_alignParentLeft	如果值为 true，则将组件定位到父容器组件的最左边
android:layout_alignParentRight	如果值为 true，则将组件定位到父容器组件的最右边
android:layout_alignParentTop	如果值为 true，则将组件定位到父容器组件的最上方
android:layout_centerInParent	如果值为 true，则将组件定位到父容器组件的中央
android:layout_below	将组件定位到指定组件的下方

下面举例说明 RelativeLayout 容器组件的使用方法。在 Android Studio 中新建一个名为 Ch0505 的应用程序工程，然后修改布局文件 res/layout/activity_main.xml 中的代码，修改后的代码如下：

```xml
<?xml version="1.0" encoding="utf-8"?>
<RelativeLayout xmlns:android="http://schemas.android.com/apk/res/android"
    android:layout_width="match_parent"
    android:layout_height="match_parent">

    <TextView
        android:id="@+id/userNameLbl"
        android:layout_width="match_parent"
        android:layout_height="wrap_content"
        android:layout_alignParentTop="true"
        android:text="@string/username" />

    <EditText
        android:id="@+id/userNameText"
        android:layout_width="match_parent"
        android:layout_height="wrap_content"
        android:hint=""
        android:layout_below="@id/userNameLbl" />

    <TextView
        android:id="@+id/pwdLbl"
        android:layout_width="match_parent"
        android:layout_height="wrap_content"
        android:layout_below="@id/userNameText"
        android:text="@string/password" />

    <EditText
        android:id="@+id/pwdText"
        android:layout_width="match_parent"
```

```
            android:layout_height="wrap_content"
            android:hint=""
            android:layout_below="@id/pwdLbl" />

        <TextView
            android:id="@+id/pwdCriteria"
            android:layout_width="match_parent"
            android:layout_height="wrap_content"
            android:layout_below="@id/pwdText"
            android:text="@string/criteria" />

        <TextView
            android:id="@+id/disclaimerLbl"
            android:layout_width="match_parent"
            android:layout_height="wrap_content"
            android:layout_alignParentBottom="true"
            android:text="@string/risk" />

</RelativeLayout>
```

修改 res/values/strings.xml 文件中的代码，修改后的代码如下：

```
<?xml version="1.0" encoding="utf-8"?>
<resources>
    <string name="app_name">Ch0505</string>

    <string name="username">用户名</string>
    <string name="password">密码</string>
    <string name="criteria">密码规范</string>
    <string name="risk">风险</string>

</resources>
```

运行 Ch0505 应用程序，运行效果如图 5-9 所示。

5.4.3 FrameLayout 容器组件

FrameLayout 容器组件是一种以层叠方式进行布局的容器组件，每次只能显示其中的一个组件。FrameLayout 容器组件为布局在其中的组件提供了一个 XML 配置属性——android:layout_gravity。通过 android:layout_gravity 属性，布局在 FrameLayout 容器组件中的组件可以指定自己在 FrameLayout 容器组件中的重心位置，如靠左、靠右等。

图 5-9 Ch0505 应用程序的运行效果

举例说明 FrameLayout 容器组件的使用方法。在 Android

Studio 中新建一个名为 Ch0506 的 Android 应用程序工程，在主界面中显示 4 张图片，需要先在 mipmap 目录下存储 4 个图片资源文件，再修改布局文件 res/layout/activity_main.xml 中的代码。修改后的 activity_main.xml 文件中的代码如下：

```xml
<?xml version="1.0" encoding="utf-8"?>
<FrameLayout xmlns:android="http://schemas.android.com/apk/res/android"
    android:id="@+id/frmLayout"
    android:layout_width="match_parent"
    android:layout_height="match_parent">

    <ImageView
        android:id="@+id/id_iv01"
        android:layout_width="match_parent"
        android:layout_height="match_parent"
        android:scaleType="fitCenter"
        android:contentDescription="@string/text_empty"
        android:src="@mipmap/png0001"
        android:visibility="visible" />

    <ImageView
        android:id="@+id/id_iv02"
        android:layout_width="match_parent"
        android:layout_height="match_parent"
        android:scaleType="fitCenter"
        android:contentDescription="@string/text_empty"
        android:src="@mipmap /png0002"
        android:visibility="gone" />

    <ImageView
        android:id="@+id/id_iv03"
        android:layout_width="match_parent"
        android:layout_height="match_parent"
        android:scaleType="fitCenter"
        android:contentDescription="@string/text_empty"
        android:src="@mipmap /png0003"
        android:visibility="gone" />

    <ImageView
        android:id="@+id/id_iv04"
        android:layout_width="match_parent"
        android:layout_height="match_parent"
        android:scaleType="fitCenter"
```

```
            android:contentDescription="@string/text_empty"
            android:src="@mipmap /png0004"
            android:visibility="gone" />

</FrameLayout>
```

在 activity_main.xml 文件中，在 FrameLayout 容器组件中放置了 4 个 ImageView 组件，并且只显示第一个 ImageView 组件，将其他 ImageView 组件的 android:visibility 属性值设置为 gone，也可以将该属性值设置为 invisible。将 android:visibility 属性值设置为 gone 与设置为 invisible 的相同点是，它们都让组件不再显示；不同点是，如果将 android:visibility 属性值设置为 gone，那么不但不显示组件，而且组件不占据显示空间。

修改 res/values/strings.xml 文件中的代码，修改后的代码如下：

```
<?xml version="1.0" encoding="utf-8"?>
<resources>
    <string name="app_name">Ch0506</string>
    <string name="text_empty"></string>
</resources>
```

在 strings.xml 文件中定义了布局文件中引用的字符串资源。

修改 MainActivity.java 文件中的代码，修改后的代码如下：

```java
package com.example.ch0506;

import androidx.appcompat.app.AppCompatActivity;

import android.os.Bundle;
import android.view.View;
import android.widget.ImageView;

public class MainActivity extends AppCompatActivity
                                    implements View.OnClickListener {

    private ImageView iv01,iv02,iv03,iv04;

    @Override
    protected void onCreate(Bundle savedInstanceState) {
        super.onCreate(savedInstanceState);
        setContentView(R.layout.activity_main);

        iv01 = this.findViewById(R.id.id_iv01);
        iv01.setOnClickListener(this);
        iv02 = this.findViewById(R.id.id_iv02);
        iv02.setOnClickListener(this);
        iv03 = this.findViewById(R.id.id_iv03);
```

```
            iv03.setOnClickListener(this);
            iv04 = this.findViewById(R.id.id_iv04);
            iv04.setOnClickListener(this);
        }

        public void onClick(View v) {
            int id = v.getId();
            switch(id) {
                case R.id.id_iv01:
                    iv01.setVisibility(View.GONE);
                    iv02.setVisibility(View.VISIBLE);
                    break;

                case R.id.id_iv02:
                    iv02.setVisibility(View.GONE);
                    iv03.setVisibility(View.VISIBLE);
                    break;

                case R.id.id_iv03:
                    iv03.setVisibility(View.GONE);
                    iv04.setVisibility(View.VISIBLE);
                    break;

                case R.id.id_iv04:
                    iv04.setVisibility(View.GONE);
                    iv01.setVisibility(View.VISIBLE);
                    break;
            }
        }
    }
```

在 MainActivity 类的 onCreate()方法中，首先获取各个 ImageView 组件的引用，并且监听对它们的点击事件。然后，在 OnClickListener 接口的 onClick()方法中，判断哪一个 ImageView 组件被点击了，并且根据当前被点击 ImageView 组件的 id 显示下一个 ImageView 组件。运行 Ch0506 应用程序，会从第 2 张图片开始显示，并且一直循环显示 4 张图片。

5.4.4 ScrollView 容器组件

ScrollView 容器组件是 FrameLayout 容器组件的子组件，主要用于提供垂直滚动条，进而将超出物理屏幕的内容显示出来。

在一般情况下，可以将一个采用垂直布局方式的 LinearLayout 容器组件作为 ScrollLayout 容器组件的子组件，在 LinearLayout 容器组件中显示超出屏幕物理高度的内容。

下面举例说明 ScrollView 容器组件的使用方法。在 Android Studio 中新建一个名为 Ch0507

的 Android 应用程序工程，修改布局文件 res/layout/activity_main.xml 中的代码。虽然 activity_main.xml 文件中的代码看起来有点长，但它的结构很简单，就是在 ScrollView 容器组件中嵌套一个 LinearLayout 容器组件，再在这个 LinearLayout 容器组件中嵌套多个 LinearLayout 容器组件，在每个二级嵌套的 LinearLayout 容器组件中显示一种颜色及其编码。activity_main.xml 文件中的代码如下：

```xml
<?xml version="1.0" encoding="utf-8"?>
<ScrollView xmlns:android="http://schemas.android.com/apk/res/android"
    android:layout_width="match_parent"
    android:layout_height="wrap_content">

    <LinearLayout
        android:layout_width="match_parent"
        android:layout_height="wrap_content"
        android:orientation="vertical" >

        <LinearLayout
            android:layout_width="match_parent"
            android:layout_height="wrap_content"
            android:orientation="horizontal" >

            <View
                android:layout_width="0dp"
                android:layout_height="120dp"
                android:layout_weight="4"
                android:background="#000000" />

            <TextView
                android:layout_width="0dp"
                android:layout_height="120dp"
                android:layout_weight="2"
                android:gravity="center"
                android:text="@string/text_000000" />
        </LinearLayout>

        <LinearLayout
            android:layout_width="match_parent"
            android:layout_height="wrap_content"
            android:orientation="horizontal" >

            <View
```

```xml
            android:layout_width="0dp"
            android:layout_height="120dp"
            android:layout_weight="4"
            android:background="#440000" />

        <TextView
            android:layout_width="0dp"
            android:layout_height="120dp"
            android:layout_weight="2"
            android:gravity="center"
            android:text="@string/text_440000" />
    </LinearLayout>

    <LinearLayout
        android:layout_width="match_parent"
        android:layout_height="wrap_content"
        android:orientation="horizontal" >

        <View
            android:layout_width="0dp"
            android:layout_height="120dp"
            android:layout_weight="4"
            android:background="#884400" />

        <TextView
            android:layout_width="0dp"
            android:layout_height="120dp"
            android:layout_weight="2"
            android:gravity="center"
            android:text="@string/text_884400" />
    </LinearLayout>

    <LinearLayout
        android:layout_width="match_parent"
        android:layout_height="wrap_content"
        android:orientation="horizontal" >

        <View
            android:layout_width="0dp"
            android:layout_height="120dp"
```

```xml
        android:layout_weight="4"
        android:background="#aa8844" />

    <TextView
        android:layout_width="0dp"
        android:layout_height="120dp"
        android:layout_weight="2"
        android:gravity="center"
        android:text="@string/text_aa8844" />
</LinearLayout>

<LinearLayout
    android:layout_width="match_parent"
    android:layout_height="wrap_content"
    android:orientation="horizontal" >

    <View
        android:layout_width="0dp"
        android:layout_height="120dp"
        android:layout_weight="4"
        android:background="#ffaa88" />

    <TextView
        android:layout_width="0dp"
        android:layout_height="120dp"
        android:layout_weight="2"
        android:gravity="center"
        android:text="@string/text_ffaa88" />
</LinearLayout>

<LinearLayout
    android:layout_width="match_parent"
    android:layout_height="wrap_content"
    android:orientation="horizontal" >

    <View
        android:layout_width="0dp"
        android:layout_height="120dp"
        android:layout_weight="4"
        android:background="#ffffaa" />
```

```xml
        <TextView
            android:layout_width="0dp"
            android:layout_height="120dp"
            android:layout_weight="2"
            android:gravity="center"
            android:text="@string/text_ffffaa" />
    </LinearLayout>

    <LinearLayout
        android:layout_width="match_parent"
        android:layout_height="wrap_content"
        android:orientation="horizontal" >

        <View
            android:layout_width="0dp"
            android:layout_height="120dp"
            android:layout_weight="4"
            android:background="#ffffff" />

        <TextView
            android:layout_width="0dp"
            android:layout_height="120dp"
            android:layout_weight="2"
            android:gravity="center"
            android:text="@string/text_ffffff" />
    </LinearLayout>
</LinearLayout>
</ScrollView>
```

修改 res/values/strings.xml 文件中的代码，在其中定义布局文件中引用的字符串资源，代码如下：

```xml
<?xml version="1.0" encoding="utf-8"?>
<resources>

    <string name="app_name">Ch0507</string>

    <string name="text_000000">#000000</string>
    <string name="text_440000">#440000</string>
    <string name="text_884400">#884400</string>
```

```xml
    <string name="text_aa8844">#aa8844</string>
    <string name="text_ffaa88">#ffaa88</string>
    <string name="text_ffffaa">#ffffaa</string>
    <string name="text_ffffff">#ffffff</string>
```

```
</resources>
```

运行 Ch0507 应用程序，运行效果如图 5-10 所示。由于显示的内容超出了物理屏幕的高度，因此自动出现了垂直滚动条。

图 5-10　Ch0507 应用程序的运行效果

> **注意**

ScrollView 容器组件只可以实现垂直滚动功能。如果要实现水平滚动功能，则可以使用 Android 提供的 HorizontalScrollView 容器组件。HorizontalScrollView 容器组件的使用方法与 ScrollView 容器组件的使用方法类似。

5.4.5　ConstraintLayout 容器组件

ConstraintLayout 容器组件是一种采用约束布局的容器组件，是在创建 Android 应用程序工程时使用的默认容器组件。ConstraintLayout 容器组件采用相对位置对新的组件进行布局，这一点与前面介绍的 RelativeLayout 容器组件类似，但 ConstraintLayout 容器组件提供了更多的 XML 布局属性。ConstraintLayout 容器组件常用的 XML 布局属性如表 5-3 所示。

表 5-3　ConstraintLayout 容器组件常用的 XML 布局属性

类　　别	属　性　名	含　　义
相对位置	layout_constraintLeft_toLeftOf layout_constraintLeft_toRightOf layout_constraintRight_toLeftOf layout_constraintRight_toRightOf layout_constraintTop_toTopOf layout_constraintTop_toBottomOf layout_constraintBottom_toTopOf layout_constraintBottom_toBottomOf layout_constraintBaseline_toBaselineOf layout_constraintStart_toEndOf layout_constraintStart_toStartOf layout_constraintEnd_toStartOf layout_constraintEnd_toEndOf	组件 B 相对于组件 A 的位置。例如，以下代码表示将组件 B 的左侧与组件 A 的右侧对齐，也就是将组件 B 放置在组件 A 的右侧。 <Button android:id="@+id/buttonA" ... /> <Button android:id="@+id/buttonB" ... app:layout_constraintLeft_toRightOf= "@+id/buttonA" />
外边距	android:layout_marginStart android:layout_marginEnd android:layout_marginLeft android:layout_marginTop android:layout_marginRight android:layout_marginBottom	设置组件 B 相对于组件 A 的外边距。例如，以下代码表示组件 A 和组件 B 占满父容器组件的宽度，并且将组件 B 定位到组件 A 的下方，他们之间的距离为 100dp。 <TextView 　　android:id="@+id/tv01" 　　android:layout_width="wrap_content" 　　android:layout_height="wrap_content" 　　android:text="Hello World!" 　　app:layout_constraintLeft_toLeftOf="parent" 　　app:layout_constraintRight_toRightOf="parent" 　　app:layout_constraintTop_toTopOf="parent" /> <TextView 　　android:id="@+id/tv02" 　　android:layout_width="wrap_content" 　　android:layout_height="wrap_content" 　　android:text="Hello Me!" 　　app:layout_constraintLeft_toLeftOf="parent" 　　app:layout_constraintRight_toRightOf="parent" 　　app:layout_constraintTop_toBottomOf="@id/tv01" 　　android:layout_marginTop="100dp"/>
居中并设置位置偏置	layout_constraintHorizontal_bias layout_constraintVertical_bias	设置组件 B 相对于组件 A 的水平或垂直方向的位置偏置。例如，以下代码表示设置 TextView 组件采用水平布局方式，但是向左偏置 30%。 <TextView 　　android:id="@+id/tv01" 　　android:layout_width="wrap_content" 　　android:layout_height="wrap_content" 　　android:text="Hello World!" 　　app:layout_constraintLeft_toLeftOf="parent" 　　app:layout_constraintRight_toRightOf="parent" 　　app:layout_constraintHorizontal_bias="0.3" 　　app:layout_constraintTop_toTopOf="parent" />

续表

类别	属性名	含义

弧形定位	layout_constraintCircle
	layout_constraintCircleRadius
	layout_constraintCircleAngle
		定位组件 B 相对于组件 A 的位置，可以通过弧形的半径和角度定位组件 B 相对于组件 A 的位置。例如，以下代码表示将组件 B 定位在组件 A 半径为 100dp、角度为 45 度的位置。
		\<Button android:id="@+id/buttonA" ... /\>
		\<Button android:id="@+id/buttonB" ...
		    app:layout_constraintCircle="@+id/buttonA"
		    app:layout_constraintCircleRadius="100dp"
		    app:layout_constraintCircleAngle="45" /\>

尺寸约束	android:minWidth android:minHeight android:maxWidth android:maxHeight	在将 ConstraintLayout 容器组件的宽度和高度都设置为"wrap_content"时，可以使用这些属性设置组件的最小尺寸和最大尺寸

下面举例说明 ConstraintLayout 容器组件的使用方法。在 Android Studio 中新建一个名为 Ch0508 的 Android 应用程序工程，修改布局文件 res/layout/activity_main.xml 中的代码，修改后的代码如下：

```xml
<?xml version="1.0" encoding="utf-8"?>
<androidx.constraintlayout.widget.ConstraintLayout
    xmlns:android="http://schemas.android.com/apk/res/android"
    xmlns:app="http://schemas.android.com/apk/res-auto"
    xmlns:tools="http://schemas.android.com/tools"
    android:layout_width="match_parent"
    android:layout_height="match_parent"
    tools:context=".MainActivity">

    <TextView
        android:id="@+id/tv01"
        android:layout_width="wrap_content"
        android:layout_height="wrap_content"
        android:text="Hello World!"
        app:layout_constraintLeft_toLeftOf="parent"
        app:layout_constraintRight_toRightOf="parent"
        app:layout_constraintTop_toTopOf="parent"
        app:layout_constraintBottom_toBottomOf="parent"/>

    <TextView
        android:id="@+id/tv02"
        android:layout_width="wrap_content"
        android:layout_height="wrap_content"
        android:text="Hello Me!"
        app:layout_constraintCircle="@id/tv01"
```

```
            app:layout_constraintCircleRadius="100dp"
            app:layout_constraintCircleAngle="45" />
</androidx.constraintlayout.widget.ConstraintLayout>
```

在上述布局文件中，我们使 id 为 tv01 的 TextView 组件占满整个父容器组件，然后，将 id 为 tv02 的 TextView 组件相对于 id 为 tv01 的 TextView 组件进行弧形定位：将 id 为 tv02 的 TextView 组件定位在 id 为 tv01 的 TextView 组件半径为 100dp、角度为 45 度的位置。

运行 Ch0508 应用程序，运行效果如图 5-11 所示。

图 5-11　Ch0508 应用程序的运行效果

5.5　同步练习二

编写一个简单的计算器应用程序，完成基本的加、减、乘、除运算，并且界面布局简洁、美观。

5.6　AdapterView 组件

在 Android 应用程序开发中，AdapterView 组件是一类常用且非常重要的组件。以列表的形式显示信息的 ListView 组件、以网格形式浏览照片缩略图的 GridView 组件、以下拉列表形式显示可选项的 Spinner 组件等都是 AdapterView 组件的子组件。下面对 AdapterView 组件进行详细介绍。

5.6.1 AdapterView 组件入门

在介绍 AdapterView 组件的使用方法前，我们先通过一个案例讲解 AdapterView 组件的编程模式。使用 ListView 组件以列表的形式显示 Android 平台上已经安装的所有应用程序信息，这些信息包括应用程序的图标、名称和入口 Activity 的类名，当显示的内容超出物理屏幕可用区域时，可以进行滚动，如图 5-12 所示。

在图 5-12 中，屏幕中显示的部分是一个 ListView 组件，小框框住的部分是列表中的一个列表项。在应用程序中使用 ListView 组件显示信息，必须完成以下的工作。

（1）在界面布局中包含一个 ListView 组件。
（2）对列表中的列表项进行布局。
（3）设计一个实现了 Adapter 接口的类，用于为 ListView 组件提供需要显示的数据。

基于前面对 UI 组件的介绍，将 ListView 组件纳入布局，并且对列表项进行布局不是一件难事，但是要用好 ListView 组件或其他 AdapterView 组件，我们需要对 Adapter 接口有一个深入的理解。

图 5-12　Android 平台上已经安装的应用程序信息

5.6.2 Adapter 接口

前面提到过，ListView（列表）组件、GridView（网格）组件和 Spinner（下拉列表）组件都是 AdapterView 组件的子组件，这些组件只负责显示数据，这些要显示的数据必须通过 Adapter 接口的实现类进行管理。以使用 ListView 组件显示数据为例，说明 AdapterView 组件与 Adapter 接口的关系，如图 5-13 所示。

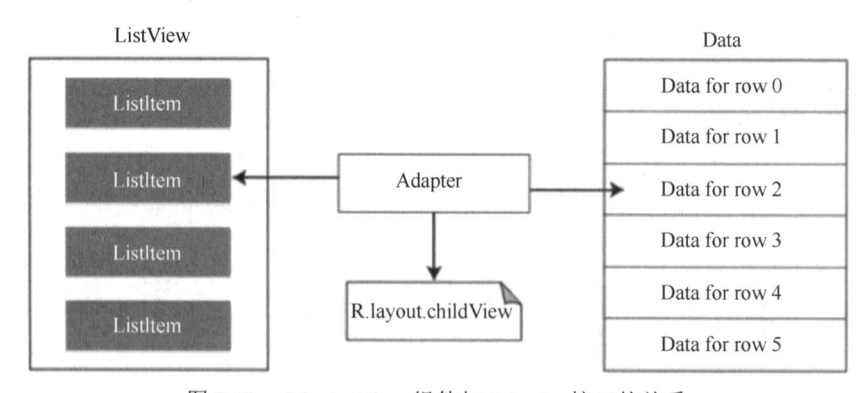

图 5-13　AdapterView 组件与 Adapter 接口的关系

根据图 5-13 可知，ListView 组件主要负责显示数据，并且对滚动、点击列表项等事件进行处理，要显示的数据由 Adapter 接口提供。其实，Adapter 接口不仅要提供数据，还要对显示在 ListView 组件中的数据进行布局封装，使其形成完整的组件，并且将其作为一个列表项提供给 ListView 组件，然后将这个组件显示出来。因此，要在 AdapterView 组件中显示数据，大部分工作都是在 Adapter 接口中完成的。

那么，AdapterView 组件是如何与 Adapter 接口进行交互，从而实现信息显示功能的呢？AdapterView 组件显示数据的处理逻辑如下。

（1）当 AdapterView 组件要显示一项数据时，它首先会调用 Adapter 接口的 getView() 方法，并且传递一个要显示数据的位置参数。

（2）Adapter 接口根据这个位置参数从 Data 中获取指定的数据，并且根据 R.layout.childView.xml 布局文件的布局样式将数据填入样式布局，然后将构建好的 View 返回给 AdapterView 组件。

（3）AdapterView 组件将返回的 View 作为子 View 显示在 View 组件中。

在 Android 帮助文档中，对 Adapter 接口的描述如下：Adapter 接口对象是 AdapterView 组件和要显示的数据之间的桥梁，它提供了对要显示的数据的访问机制，同时，Adapter 接口对象还必须为 AdapterView 组件生成要显示的组件，并且将其提供给 AdapterView 组件进行显示。

这个描述可能有点抽象，没有关系，我们会通过案例介绍如何定义实现了 Adapter 接口的类。在这之前，我们需要了解 Adapter 接口中的方法及其含义。Adapter 接口中的常用方法及其含义如表 5-4 所示。

表 5-4　Adapter 接口中的常用方法及其含义

方 法 名 称	含　　义
int getCount( )	返回要显示的数据集中的数据总数
Object getItem(int position)	返回要显示的数据集中指定位置的数据对象
long getItemId(int position)	返回要显示的数据集中指定位置的数据 id
View getView(int position, View convertView, ViewGroup parent)	将指定位置的数据构建成一个可以显示在 AdapterView 组件中的组件，并且将其返回给 AdapterView 组件进行显示

为了方便进行 Android 应用程序设计，Android 平台的 SDK 提供了几个实现 Adapter 接口的基础类，它们分别是 BaseAdapter 类、ArrayAdapter 类、SimpleAdapter 类、SimpleCursorAdapter 类。在一般情况下，我们需要对这些基础类进行扩展，从而实现特殊功能。

5.6.3　ListView 组件

1．ListView 组件的基础

ListView 组件以垂直布局方式显示数据列表，当要显示的数据较多，总高度超过屏幕高度时，能以垂直滚动的方式显示其他数据。

下面来看一个案例：ListView 组件以列表的形式显示书的名称，该案例应用程序的运行效果如图 5-14 所示。通过该案例，介绍 ListView 组件及 Adapter 接口的基本使用方法。

在 Android Studio 中新建一个名为 Ch0509 的 Android 应用程序工程。下面按照 5.6.1 节介绍的使用 AdapterView 组件显示信息的 3 个步骤完成工作。

图 5-14 使用 ListView 组件的案例应用程序的运行效果

（1）在界面布局中包含一个 ListView 组件。

主界面布局文件是 activity_main.xml，修改该文件中的代码，修改后的代码如下：

```xml
<?xml version="1.0" encoding="utf-8"?>
<RelativeLayout xmlns:android="http://schemas.android.com/apk/res/android"
    android:layout_width="match_parent"
    android:layout_height="wrap_content">

    <ListView
        android:id="@+id/id_lv"
        android:layout_width="match_parent"
        android:layout_height="match_parent"/>

</RelativeLayout>
```

在主界面布局文件中，RelativeLayout 容器组件中包含一个 ListView 组件，并且这个 ListView 组件在水平方向上占满整个显示空间，在垂直方向上按内容多少占用显示空间。

（2）对列表中的列表项进行布局。

右击工程的 res/layout 目录，在弹出的快捷菜单中选择 New→Layout Resource File 命令，如图 5-15 所示。在弹出的对话框的 File name 文本框中输入 "list_item.xml"，单击 Finish 按钮，创建列表项布局文件 list_item.xml。修改 list_item.xml 文件中的代码，修改后的代码如下：

第 5 章 Android 中常用的 UI 组件

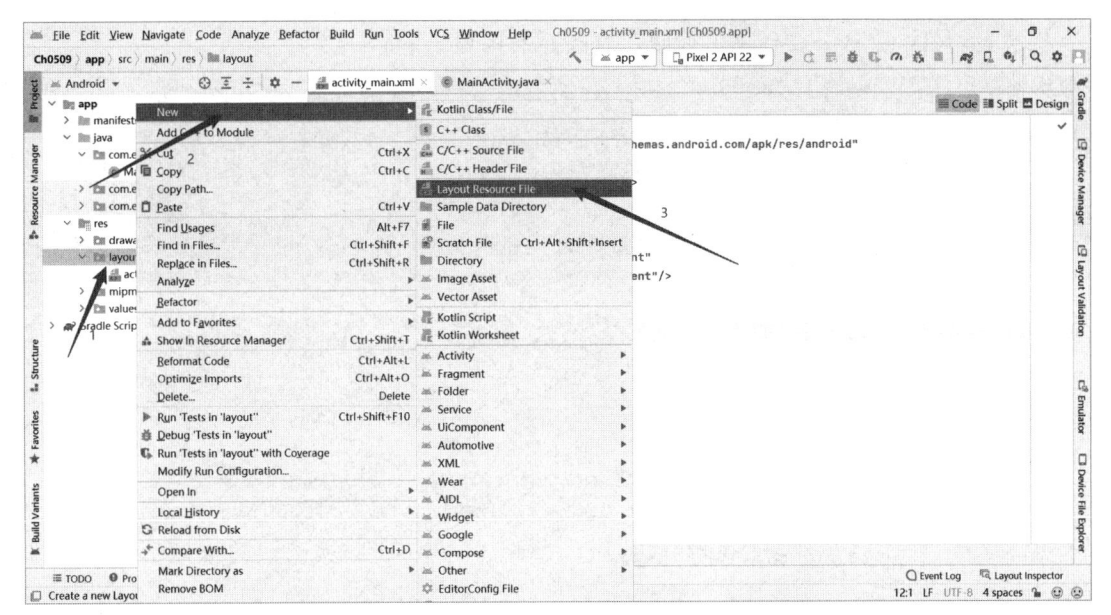

图 5-15 新建一个布局文件

```xml
<?xml version="1.0" encoding="utf-8"?>
<TextView xmlns:android="http://schemas.android.com/apk/res/android"
    android:id="@+id/id_book_name"
    android:layout_width="match_parent"
    android:layout_height="match_parent"
/>
```

这个布局文件是列表中列表项的布局文件：每个列表项都使用 TextView 组件显示书的名称。

（3）定义一个实现了 Adapter 接口的类，用于为 ListView 组件提供要显示的数据。

在 Ch0509 应用程序工程中，新建一个名为 MyAdapter 的 Java 类，操作步骤如图 5-16 所示。

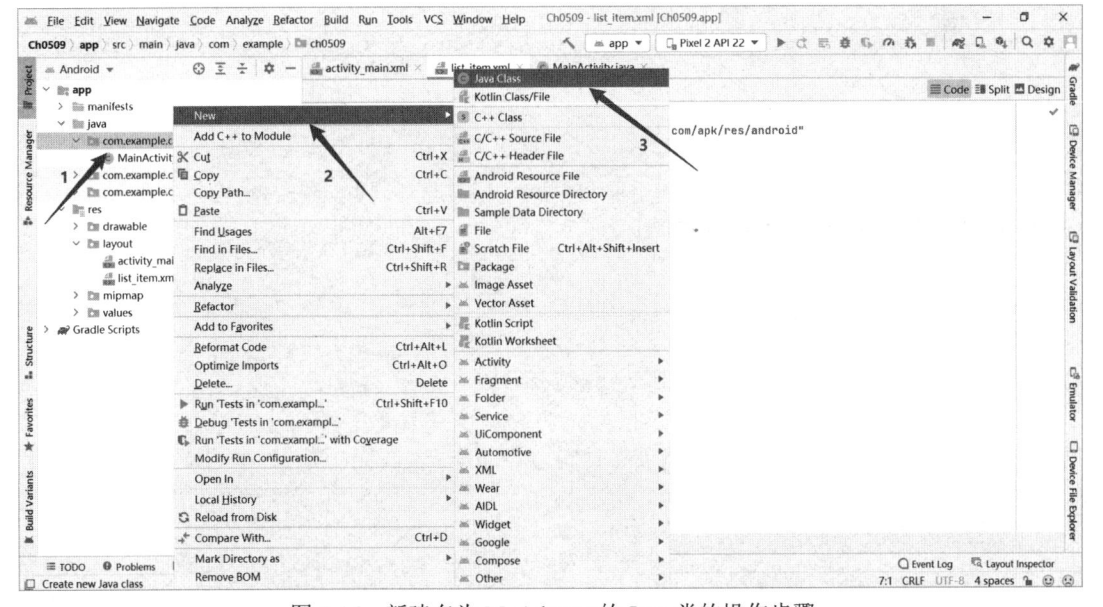

图 5-16 新建名为 MyAdapter 的 Java 类的操作步骤

修改 MyAdapter.java 文件中的代码，修改后的代码如下：

```java
package com.example.ch0509;

import android.annotation.SuppressLint;
import android.content.Context;
import android.view.LayoutInflater;
import android.view.View;
import android.view.ViewGroup;
import android.widget.BaseAdapter;
import android.widget.TextView;

public class MyAdapter extends BaseAdapter {
    private final String[] books = {"Java 程序设计", "Android 应用开发",
            "Oracle 数据库管理指南","Java Web 程序设计", "软件工程之系统工程师之路"};

    LayoutInflater inflater;
    int id_list_item;

    public MyAdapter(Context context, int id_list_item) {
        this.id_list_item = id_list_item;
        inflater = (LayoutInflater)
                context.getSystemService(Context.LAYOUT_INFLATER_SERVICE);
    }

    @Override
    public int getCount() {
        return books.length;
    }

    @Override
    public Object getItem(int position) {
        return books[position];
    }

    @Override
    public long getItemId(int position) {
        return position;
    }

    @SuppressLint("ViewHolder")
    @Override
    public View getView(int position, View convertView, ViewGroup parent) {
```

```
        TextView tv;
        tv = (TextView)inflater.inflate(id_list_item, parent, false);
        tv.setText(books[position]);

        return tv;
    }
}
```

如前所述，编写自己的 Adapter 是使用 AdapterView 组件的关键。为了在 ListView 组件中显示书的名称，我们创建了 Adapter 接口的实现类 MyAdapter。为了简化设计，使 MyAdapter 类直接继承 BaseAdapter 类。需要注意的是，BaseAdapter 类只是 Adapter 接口的简单实现类，它并没有做什么工作，根据需要，我们重载了其中的几个重要方法，下面重点介绍 getCount() 和 getView() 方法的具体实现。

当 ListView 组件要显示数据时，它会先调用 MyAdapter 类的 getCount() 方法，用于询问 MyAdapter 类中共有多少条数据需要显示，以便 ListView 组件做一些准备工作。MyAdapter 类会将 books 数组的长度返回给 ListView 组件。然后系统会调用 MyAdapter 类的 getView() 方法，用于获得要显示的数据。因此，在 getView() 方法中，需要先使用布局展开器将列表项布局资源 R.layout.list_item 展开成视图。因为 list_item.xml 文件中只有一个 TextView 组件，所以可以将其强制转换成 TextView 对象。然后将 books 数组中书的名称显示在该 TextView 组件中，并且将这个组装好的 TextView 组件返回给 ListView 组件，使 ListView 组件在列表中显示该 TextView 组件。

在完成以上工作后，修改 MainActivity.java 文件中的代码，在其中的 onCreate() 方法中获取主界面中的 ListView 组件，并且告诉这个 ListView 组件从 MyAdapter 对象中获取要显示的数据。修改后的 MainActivity.java 文件中的代码如下：

```
package com.example.ch0509;

import androidx.appcompat.app.AppCompatActivity;

import android.os.Bundle;
import android.widget.ListView;

public class MainActivity extends AppCompatActivity {

    @Override
    protected void onCreate(Bundle savedInstanceState) {
        super.onCreate(savedInstanceState);
        setContentView(R.layout.activity_main);

        ListView lv = this.findViewById(R.id.id_lv);   //获取ListView组件的引用
        MyAdapter mad = new MyAdapter(this, R.layout.list_item);
        //告诉ListView组件从MyAdapter对象mad中获取数据
```

```
            lv.setAdapter(mad);
        }
    }
```

注意 onCreate()方法中的以下代码。

```
        ListView lv = (ListView)this.findViewById(R.id.id_lv);  //获取ListView组件的引用
        MyAdapter mad = new MyAdapter(this, R.layout.list_item);
        lv.setAdapter(mad);   //告诉ListView组件从MyAdapter对象mad中获取数据
```

这段代码主要用于获取 ListView 组件的引用，并且告诉这个 ListView 组件从 MyAdapter 对象 mad 中获取数据。

运行 Ch0509 应用程序，即可得到图 5-14 所示的运行效果。

2．使 ListView 组件显示的信息更漂亮

使用 ListView 组件显示书的名称，这种效果太单调了。下面修改 Ch0509 应用程序，使 ListView 组件中的每一行都显示书的名称和封面图片，如图 5-17 所示。

图 5-17　每一行都显示书的名称和封面图片

为了实现这种显示效果，我们需要修改列表项布局文件 res/layout/list_item.xml 和 MyAdapter.java 文件。

首先修改列表项布局文件 list_item.xml。现在的列表项不仅需要显示书的名称的 TextView 组件，还需要显示书的封面的 ImageView 组件。修改后的 list_item.xml 文件中的代码如下：

```
<?xml version="1.0" encoding="utf-8"?>
<LinearLayout xmlns:android="http://schemas.android.com/apk/res/android"
    android:layout_width="match_parent"
    android:layout_height="match_parent"
```

```xml
    android:orientation="horizontal" >

    <ImageView
        android:id="@+id/id_book_photo"
        android:layout_width="60dp"
        android:layout_height="60dp"
        android:contentDescription="@string/text_empty"    //不要产生警告
    />

    <TextView
        android:id="@+id/id_book_name"
        android:layout_width="match_parent"
        android:layout_height="match_parent"
        android:gravity="center_vertical"
    />

</LinearLayout>
```

在列表项布局文件 list_item.xml 中添加了一个用于显示图片的 ImageView 组件。需要注意的是，为了使 ImageView 组件不产生警告，在 ImageView 组件的 android:contentDescription 属性中赋予了一个 text_empty 字符串资源的引用。因此，需要在 res/values/strings.xml 文件中定义相应的字符串资源。修改后的 strings.xml 文件中的代码如下：

```xml
<?xml version="1.0" encoding="utf-8"?>
<resources>

    <string name="app_name">Ex05ListView01</string>
    <string name="text_empty"></string>

</resources>
```

修改 MyAdapter.java 文件中的代码，修改后的代码如下：

```java
package com.example.ch0509;

import android.annotation.SuppressLint;
import android.content.Context;
import android.view.LayoutInflater;
import android.view.View;
import android.view.ViewGroup;
import android.widget.BaseAdapter;
import android.widget.ImageView;
import android.widget.LinearLayout;
import android.widget.TextView;

public class MyAdapter extends BaseAdapter {
```

```java
        private final BookItem[] books = {new BookItem("Java 程序设计", R.mipmap.png0001),
                new BookItem("Android 应用开发", R.mipmap.png0002),
                new BookItem("Oracle 数据库管理指南", R.mipmap.png0003),
                new BookItem("Java Web 程序设计", R.mipmap.png0004),
                new BookItem("软件工程之系统工程师之路", R.mipmap.png0005)};

        LayoutInflater inflater;
        int id_list_item;

        public MyAdapter(Context context, int id_list_item) {
            this.id_list_item = id_list_item;
            inflater = (LayoutInflater)
                    context.getSystemService(Context.LAYOUT_INFLATER_SERVICE);
        }

        @Override
        public int getCount() {
            return books.length;
        }

        @Override
        public Object getItem(int position) {
            return books[position];
        }

        @Override
        public long getItemId(int position) {
            return position;
        }

        @SuppressLint("ViewHolder") @Override
        public View getView(int position, View convertView, ViewGroup parent) {
            LinearLayout ll = (LinearLayout)inflater.inflate(id_list_item, parent, false);
            ImageView iv = ll.findViewById(R.id.id_book_photo);
            iv.setImageResource(books[position].photo);
            TextView tv;
            tv = ll.findViewById(R.id.id_book_name);
            tv.setText(books[position].name);

            return ll;
        }
```

```
    private static class BookItem {
        String name;
        int photo;

        public BookItem(String name, int photo) {
            this.name = name;
            this.photo = photo;
        }
    }
}
```

为了表示书的相关信息，我们定义了一个内部类 BookItem，其中包含书的名称和书的封面，并且在 books 数组变量中初始化书的相关信息。为了能够显示图片，我们在 res/mipmap 目录下存储书的封面图片。

在 MyAdapter 类的 getView()方法中，仍然采用相似的过程封装要显示在 ListView 组件中的列表项：首先使用布局展开器将列表项布局文件 R.layout.list_item 展开成视图，然后在相应的 ImageView 组件中显示书的封面，在 TextView 组件中显示书的名称，并且将这些封装好的组件返回给 ListView 组件。

运行修改后的 Ch0509 应用程序，即可得到图 5-17 所示的运行效果。

3．响应对 ListView 组件中列表项的点击事件

继续完善上面的案例，当用户点击列表中的某个列表项时，会弹出一个信息提示框，用于显示当前列表项中书的名称。例如，当点击 "Android 应用开发" 列表项时，显示如图 5-18 所示的信息。

要实现上述效果，需要修改 MainActivity.java 文件中的代码，在 onCreate()方法中，监听对 ListView 组件中列表项的点击事件，并且实现对该点击事件的响应方法 onItemClick()，修改后的代码如下：

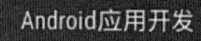

图 5-18　信息提示框

```
package com.example.ch0509;

import androidx.appcompat.app.AppCompatActivity;

import android.os.Bundle;
import android.view.View;
import android.widget.AdapterView;
import android.widget.ListView;
import android.widget.TextView;
import android.widget.Toast;

public class MainActivity extends AppCompatActivity implements AdapterView.OnItemClickListener {
    @Override
    protected void onCreate(Bundle savedInstanceState) {
```

```
        super.onCreate(savedInstanceState);
        setContentView(R.layout.activity_main);

        ListView lv = this.findViewById(R.id.id_lv);  //获取ListView组件的引用
        MyAdapter mad = new MyAdapter(this, R.layout.list_item);
        lv.setAdapter(mad);      //告诉ListView组件从MyAdapter对象mad中获取数据
        lv.setOnItemClickListener(this);
    }

    @Override
    public void onItemClick(AdapterView<?> adapterView, View view, int i, long l) {
        TextView tv = view.findViewById(R.id.id_book_name);
        String name = (String) tv.getText();
        Toast.makeText(this, name, Toast.LENGTH_LONG).show();
    }
}
```

在上述代码中，MainActivity类实现了OnItemClickListener接口，代码如下：

```
public class MainActivity extends AppCompatActivity implements AdapterView.OnItemClickListener {
```

在MainActivity类的onCreate()方法中，监听对ListView组件中列表项的点击事件，代码如下：

```
        lv.setOnItemClickListener(this);
```

在OnItemClickListener接口的onItemClick()方法的具体实现方法中，显示被点击的列表项信息，代码如下：

```
    @Override
    public void onItemClick(AdapterView<?> parent, View view, int position,
            long id) {
        TextView tv = view.findViewById(R.id.id_book_name);
        String name = (String) tv.getText();
        Toast.makeText(this, name, Toast.LENGTH_LONG).show();
    }
```

注意onItemClick()方法传进来的参数，其中的View view是被点击的对象，在本案例中，它是一个LinearLayout容器组件。因此，我们可以从view对象中找到显示书的名称的TextView组件，并且从这个TextView组件中获取书的名称。

Toast是一个经常用于显示信息的Android组件，可以使用Toast组件的makeText()静态方法构造Toast对象，该对象中包含要显示的信息，然后调用Toast组件的show()方法将所构建的Toast对象显示出来。

4．使用ListView组件显示Android平台上已安装的所有应用程序

现在实现5.6.1节中的案例，用Android的ListView组件以列表的形式显示Android平台上已经安装的所有应用程序的图标、名称和入口Activity的类名。

第 5 章 Android 中常用的 UI 组件

在 Android Studio 中新建一个名为 Ch0510 的 Android 应用程序工程。修改布局文件 activity_main.xml 中的代码，在一个 LinearLayout 容器组件中包含一个 ListView 组件，修改后的代码如下：

```xml
<?xml version="1.0" encoding="utf-8"?>
<LinearLayout xmlns:android="http://schemas.android.com/apk/res/android"
    android:layout_width="match_parent"
    android:layout_height="match_parent">

    <ListView
        android:id="@+id/lv"
        android:layout_width="fill_parent"
        android:layout_height="wrap_content"/>

</LinearLayout>
```

创建列表项布局文件 list_item.xml，修改该文件中的代码，修改后的代码如下：

```xml
<?xml version="1.0" encoding="utf-8"?>
<LinearLayout xmlns:android="http://schemas.android.com/apk/res/android"
    android:layout_width="match_parent"
    android:layout_height="match_parent"
    android:orientation="horizontal" >

    <ImageView
        android:id="@+id/id_icon"
        android:layout_width="48dp"
        android:layout_height="48dp"
        android:contentDescription="@string/text_empty" />

    <LinearLayout
        android:layout_width="match_parent"
        android:layout_height="wrap_content"
        android:orientation="vertical" >

        <TextView
            android:id="@+id/id_appName"
            android:layout_width="match_parent"
            android:layout_height="wrap_content" />

        <TextView
            android:id="@+id/id_packageName"
            android:layout_width="match_parent"
            android:layout_height="wrap_content" />
    </LinearLayout>
```

```
</LinearLayout>
```

创建 MyAdapter.java 文件,用于从 Android 平台上获取已安装的所有应用程序的图标、名称和入口 Activity 的类名,并且将其作为列表项提交给 ListView 组件显示,代码如下:

```java
package com.example.ch0510;

import android.annotation.SuppressLint;
import android.content.Context;
import android.content.pm.PackageInfo;
import android.content.pm.PackageManager;
import android.graphics.drawable.Drawable;
import android.view.LayoutInflater;
import android.view.View;
import android.view.ViewGroup;
import android.widget.BaseAdapter;
import android.widget.ImageView;
import android.widget.TextView;

import java.util.ArrayList;
import java.util.HashMap;
import java.util.List;
import java.util.Map;

public class MyAdapter extends BaseAdapter {
    ArrayList<HashMap<String, Object>> items =
            new ArrayList<HashMap<String, Object>>();
    private LayoutInflater mInflater;
    int item_layout;

    public MyAdapter(Context context, int item_layout) {
        mInflater = (LayoutInflater) context
                .getSystemService(Context.LAYOUT_INFLATER_SERVICE);
        this.item_layout = item_layout;

        // 获取 PackageManager 对象
        PackageManager pm = context.getPackageManager();
        // 获取系统安装的所有程序包的 PackageInfo 对象
        List<PackageInfo> packs = pm.getInstalledPackages(0);
        for (PackageInfo pi : packs) {
            HashMap<String, Object> map = new HashMap<String, Object>();
            map.put("icon", pi.applicationInfo.loadIcon(pm));      // 应用程序图标
            map.put("appName", pi.applicationInfo.loadLabel(pm));// 应用程序名称
```

```java
            map.put("packageName", pi.packageName);              // 包名
            // 循环读取应用程序的相关信息,并且将其存储于ArrayList对象items中
            items.add(map);
        }
    }

    @Override
    public int getCount() {
        return items.size();
    }

    @Override
    public Object getItem(int position) {
        return items.get(position);
    }

    @Override
    public long getItemId(int position) {
        return position;
    }

    @SuppressLint("ViewHolder")
    public View getView(int position, View convertView, ViewGroup parent) {
        View v;
        v = mInflater.inflate(item_layout, parent, false);

        final Map<String, ?> dataSet = items.get(position);
        if (dataSet == null) {
            return null;
        }

        ImageView iv = (ImageView)v.findViewById(R.id.id_icon);
        iv.setImageDrawable((Drawable) dataSet.get("icon"));
        TextView tv1 = (TextView)v.findViewById(R.id.id_appName);
        tv1.setText(dataSet.get("appName").toString());
        TextView tv2 = (TextView)v.findViewById(R.id.id_packageName);
        tv2.setText(dataSet.get("packageName").toString());

        return v;
    }

}
```

在 MyAdapter 类的构造方法中，首先获取一个布局展开器对象，用于在 getView()方法中展开列表项布局，然后获取一个包管理器对象，用于从 Android 平台上获取已安装的所有应用程序信息，并且将每个应用程序的信息都作为一个 map 对象存储于 ArrayList 对象中。在 getView()方法中，首先展开一个列表项布局，从 items 数组中获取相应的应用程序信息，然后将其逐个置入展开的列表项布局，最后将包装好的列表项组件返回给 ListView 组件。

修改 MainActivity.java 文件中的代码，修改后的代码如下：

```java
package com.example.ch0510;

import androidx.appcompat.app.AppCompatActivity;

import android.os.Bundle;
import android.widget.ListView;

public class MainActivity extends AppCompatActivity {

    @Override
    protected void onCreate(Bundle savedInstanceState) {
        super.onCreate(savedInstanceState);
        setContentView(R.layout.activity_main);

        ListView lv = this.findViewById(R.id.lv);    //获取 ListView 组件的引用
        MyAdapter mad = new MyAdapter(this, R.layout.list_item);
        lv.setAdapter(mad);    //告诉 ListView 组件从 MyAdapter 对象 mad 中获取数据
    }
}
```

5.6.4　Spinner 组件

Spinner 组件是 GUI 中的下拉列表。与 ListView 组件类似，我们必须为 Spinner 对象指定一个 Adapter 接口的实现类。

1. Spinner 组件的简单应用

下面构建一个简单的应用程序：在应用程序界面中显示一个下拉列表，用户可以从中选择一个选项，并且在界面中显示该选项。该案例应用程序的运行效果如图 5-19 所示。

在 Android Studio 中新建一个名为 Ch0511 的 Android 应用程序工程。修改主界面布局文件 res/layout/activity_main.xml 中的代码，修改后的代码如下：

图 5-19　使用 Spinner 组件的案例应用程序的运行效果

```xml
<?xml version="1.0" encoding="utf-8"?>
<LinearLayout xmlns:android="http://schemas.android.com/apk/res/android"
    android:layout_width="match_parent"
    android:layout_height="match_parent"
    android:orientation="vertical">

    <LinearLayout
        android:layout_width="match_parent"
        android:layout_height="wrap_content"
        android:orientation="horizontal" >

        <TextView
            android:layout_width="wrap_content"
            android:layout_height="wrap_content"
            android:text="@string/text_choice" />
        <Spinner
            android:id="@+id/id_spinner"
            android:layout_width="match_parent"
            android:layout_height="wrap_content" />

    </LinearLayout>

    <View
        android:layout_width="match_parent"
        android:layout_height="4dp"
        android:background="#aaa" />

    <TextView
        android:id="@+id/id_choice"
        android:gravity="center"
        android:layout_width="match_parent"
        android:layout_height="wrap_content" />

</LinearLayout>
```

在 activity_main.xml 文件中，在 LinearLayout 容器组件中放置了一个 LinearLayout 容器组件、一个分隔组件用的 View 组件和一个显示所选结果的 TextView 组件。在内部嵌套的 LinearLayout 容器组件中放置了一个的 TextView 组件和 Spinner 组件。

修改 res/values/strings.xml 文件中的代码，首先定义在布局文件中引用的字符串资源，然后定义一个字符串数组资源，修改后代码如下：

```xml
<?xml version="1.0" encoding="utf-8"?>
<resources>
```

```xml
    <string name="app_name">Ch0511</string>

    <string name="text_choice">选择你的所爱：</string>

    <string-array name="habit">
        <item>5 千米晨跑</item>
        <item>去健身房健身</item>
        <item>爬山</item>
        <item>2 千米游泳</item>
        <item>打篮球</item>
    </string-array>

</resources>
```

在上述代码中，定义的字符串数组资源是下拉列表中的选项，会在 MainActivity.java 文件中使用这个字符串数组资源。

修改 MainActivity.java 文件中的代码，修改后的代码如下：

```java
package com.example.ch0511;

import androidx.appcompat.app.AppCompatActivity;

import android.os.Bundle;
import android.view.View;
import android.widget.AdapterView;
import android.widget.ArrayAdapter;
import android.widget.Spinner;
import android.widget.TextView;

public class MainActivity extends AppCompatActivity implements AdapterView.OnItemSelectedListener {
    TextView choice;
    ArrayAdapter<CharSequence> adapter;

    @Override
    protected void onCreate(Bundle savedInstanceState) {
        super.onCreate(savedInstanceState);
        setContentView(R.layout.activity_main);

        choice = (TextView)this.findViewById(R.id.id_choice);

        Spinner spinner = (Spinner)this.findViewById(R.id.id_spinner);
        adapter = ArrayAdapter.createFromResource(this,
                R.array.habit, android.R.layout.simple_spinner_item);
        adapter.setDropDownViewResource(
```

```
                    android.R.layout.simple_spinner_dropdown_item);
        spinner.setAdapter(adapter);
        spinner.setOnItemSelectedListener(this);
    }

    @Override
    public void onItemSelected(AdapterView<?> parent, View view, int position,
long id) {
        choice.setText(adapter.getItem(position));
    }

    @Override
    public void onNothingSelected(AdapterView<?> parent) {
    }
}
```

在 MainActivity 类的 onCreate()方法中，首先获取 Spinner 对象 spinner，然后使用 ArrayAdapter 类的静态方法 createFromResource()创建一个 ArrayAdapter 对象 adapter，其中的 R.array.habit 参数主要用于指定一个字符串数组资源，android.R.layout.simple_spinner_item 参数主要用于指定在下拉列表未展开时的数据显示布局，在本案例中，我们使用的是 Android 平台自带的一个布局，这个布局中只有一个 TextView 组件。注意其中的代码：

```
adapter.setDropDownViewResource(
                android.R.layout.simple_spinner_dropdown_item);
```

上述代码又指定了一个布局，这个布局是下拉列表展开时的数据显示布局，我们仍然使用 Android 平台自带的一个布局,这个布局中仍然只有一个 TextView 组件。

将构建好的 adapter 对象设置为 Spinner 组件的适配器。

为了响应对下拉列表中选项的选择事件，MainActivity 类实现了 OnItemSelectedListener 接口，并且在该接口的 onItemSelected()方法的具体实现方法中，将 adapter 对象中的相应信息显示在第二个 TextView 组件中。

运行 Ch0511 应用程序,并且选择其中的一个选项，即可得到图 5-19 所示的运行效果。

2. 美化 Spinner

在下拉列表中只显示书的名称，这种效果太单调了。下面修改 Ch0511 应用程序，使下拉列表的显示效果更加美观。修改后的 Ch0511 应用程序的运行效果如图 5-20 所示。

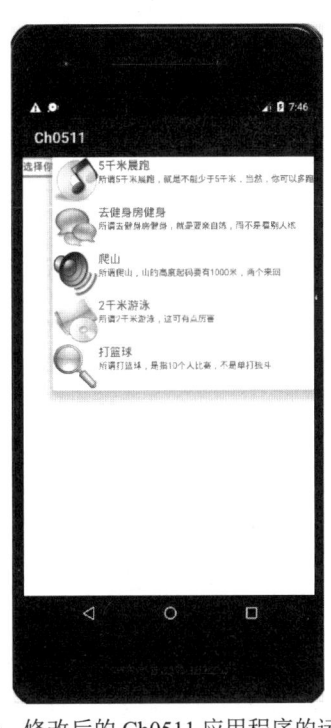

图 5-20　修改后的 Ch0511 应用程序的运行效果

首先添加几个在下拉列表中显示的图片资源文件，然后创建一个下拉列表选项布局文件 dropdown_item.xml，该文件中的代码如下：

```xml
<?xml version="1.0" encoding="utf-8"?>
<LinearLayout xmlns:android="http://schemas.android.com/apk/res/android"
    android:layout_width="match_parent"
    android:layout_height="match_parent"
    android:orientation="horizontal" >

    <ImageView
        android:id="@+id/id_imageview"
        android:layout_width="64dp"
        android:layout_height="64dp" />

    <LinearLayout
        android:layout_width="match_parent"
        android:layout_height="match_parent"
        android:orientation="vertical" >

        <TextView
            android:id="@+id/id_textview_01"
            android:textSize="16sp"
            android:layout_width="match_parent"
            android:layout_height="wrap_content" />

        <TextView
            android:id="@+id/id_textview_02"
            android:textSize="12sp"
            android:layout_width="match_parent"
            android:layout_height="wrap_content" />

    </LinearLayout>

</LinearLayout>
```

为了在选择下拉列表中的选项时显示自定义的效果，需要重新定义 ArrayAdapter 类。创建一个自定义的 MyArrayAdapter 类，并且重写其中的 getDropDownView() 方法，该方法会在用户选择下拉列表中的选项时被调用，因此，可以在该方法中创建指定的布局，并且显示指定的信息。MyArrayAdapter.java 文件的代码如下：

```java
package com.example.ch0511;

import android.content.Context;
import android.view.LayoutInflater;
import android.view.View;
```

```java
import android.view.ViewGroup;
import android.widget.ArrayAdapter;
import android.widget.ImageView;
import android.widget.TextView;

public class MyArrayAdapter extends ArrayAdapter<CharSequence> {
    private LayoutInflater mInflater;
    String[] titles;
    String[] desc = {
            "所谓5千米晨跑，就是不能少于5千米，当然，你可以多跑",
            "所谓去健身房健身，就是要亲自练，而不是看别人练",
            "所谓爬山，山的高度起码要有1000米，两个来回",
            "所谓2千米游泳，这可有点厉害",
            "所谓打篮球，是指10个人比赛，不是单打独斗"
    };
    int[] images = {R.mipmap.png0001, R.mipmap.png0002,
            R.mipmap.png0003, R.mipmap.png0004,
            R.mipmap.png0005};

    public MyArrayAdapter(Context context, int resource, CharSequence[] objects) {
        super(context, resource, objects);

        mInflater = (LayoutInflater) context
                .getSystemService(Context.LAYOUT_INFLATER_SERVICE);
        titles = context.getResources().getStringArray(R.array.habit);
    }


    @Override
    public int getCount() {
        return titles.length;
    }


    @Override
    public View getDropDownView(int position, View convertView, ViewGroup parent) {
        View v;
        v = mInflater.inflate(R.layout.dropdown_item, parent, false);

        ImageView iv = (ImageView)v.findViewById(R.id.id_imageview);
        iv.setImageResource(images[position]);
        TextView tv01 = (TextView)v.findViewById(R.id.id_textview_01);
        tv01.setText(titles[position]);
```

```
        TextView tv02 = (TextView)v.findViewById(R.id.id_textview_02);
        tv02.setText(desc[position]);

        return v;
    }
}
```

在 MyArrayAdapter 类的构造方法中,以下代码可以从资源文件中获取字符串数组资源。在 Java 代码中可以很便利地引用资源文件中定义的字符串数组资源。

```
titles = context.getResources().getStringArray(R.array.habit);
```

在 MyArrayAdapter 类的 getDropDownView()方法中,首先展开下拉列表选项布局,然后将相应的信息写入相应的组件,最后将组装好的组件返回给 Spinner 组件,以便在选择下拉列表中的选项时将其显示出来。

修改 MainActivity.java 文件中的代码,在 onCreate()方法中创建 MyArrayAdapter 对象,并且将其设置为 Spinner 组件的适配器,修改后的代码如下:

```
package com.example.ch0511;

import android.os.Bundle;
import android.view.View;
import android.widget.AdapterView;
import android.widget.ArrayAdapter;
import android.widget.Spinner;
import android.widget.TextView;

import androidx.appcompat.app.AppCompatActivity;

public class MainActivity extends AppCompatActivity  implements AdapterView.OnItemSelectedListener {
    TextView choice;
    ArrayAdapter<CharSequence> adapter;

    @Override
    protected void onCreate(Bundle savedInstanceState) {
        super.onCreate(savedInstanceState);
        setContentView(R.layout.activity_main);

        choice = (TextView)this.findViewById(R.id.id_choice);

        Spinner spinner = (Spinner)this.findViewById(R.id.id_spinner);
        adapter = new MyArrayAdapter(this, android.R.layout.simple_spinner_item,
                this.getResources().getTextArray(R.array.habit));
        spinner.setAdapter(adapter);
```

```java
            spinner.setOnItemSelectedListener(this);
        }

        @Override
        public void onItemSelected(AdapterView<?> parent, View view, int position,
                            long id) {
            choice.setText(adapter.getItem(position));
        }

        @Override
        public void onNothingSelected(AdapterView<?> parent) {
        }

}
```

运行修改后的 Ch0511 应用程序，即可得到图 5-20 所示的运行效果。

5.6.5 GridView 组件

GridView 组件能以二维表格的方式显示数据，如果数据比较多，那么该组件可以提供垂直滚动条。下面举例说明 GridView 组件的使用方法：使用 GridView 组件显示一组图片的缩略图，当点击某个缩略图时，使用 Toast 组件显示这个图片的信息，该案例应用程序的运行效果如图 5-21 所示。

图 5-21　使用 GridView 组件的案例应用程序的运行效果

在 Android Studio 中新建一个名为 Ch0512 的 Android 应用程序工程，并且在 res/mipmap 目录下存储需要显示的图片资源文件。修改布局文件 res/layout/activity_main.xml 中的代码，修改后的代码如下：

```xml
<?xml version="1.0" encoding="utf-8"?>
<RelativeLayout xmlns:android="http://schemas.android.com/apk/res/android"
    android:layout_width="match_parent"
    android:layout_height="match_parent">

    <GridView
        android:id="@+id/gridview"
        android:layout_width="match_parent"
        android:layout_height="match_parent"
        android:gravity="center"
        android:columnWidth="90dp"
        android:numColumns="auto_fit"
        android:stretchMode="columnWidth"
        android:horizontalSpacing="10dp"
        android:verticalSpacing="10dp" />

</RelativeLayout>
```

这个布局文件比较简单，在一个 RelativeLayout 容器组件中放置了一个 GridView 组件。GridView 组件的关键配置属性如下。
- android:columnWidth。用于指定 GridView 组件中每列的宽度，需要设置一个宽度值。
- android:numColumns。用于指定 GridView 组件中每行的列数，其值可以是一个带像素单位的整数值，也可以是"auto_fit"，即自动适应。
- android:stretchMode。当屏幕的宽度值不是 android:columnWidth 属性值的整数倍时，用于指定多余显示空间的分配方式，其值如下。
  - none：不进行任何扩展操作。
  - spacingWidth：将列与列之间的间隔放大。
  - columnWidth：将每列都自动放大。
  - spacingWidthUniform：在列与列之间均匀分配多余的空间。
- android:horizontalSpacing。用于指定 GridView 组件中行与行之间的分隔高度，其值为一个带像素单位的整数值。
- android:verticalSpacing。用于指定 GridView 组件中列与列之间的分隔宽度，其值为一个带像素单位的整数值。

由于 GridView 组件是 AdapterView 组件的子组件，因此，需要创建一个 Adapter 接口的实现类，用于提供要显示的数据。创建 Adapter 接口的实现类 ImageAdapter，将需要显示的图片资源文件存储于 mThumbIds 数组中，并且重写 getView()方法。ImageAdapter.java 文件中的代码如下：

```
package com.example.ch0512;

import android.content.Context;
import android.view.View;
import android.view.ViewGroup;
```

```java
import android.widget.BaseAdapter;
import android.widget.GridView;
import android.widget.ImageView;

public class ImageAdapter extends BaseAdapter {
    private final Context mContext;

    public ImageAdapter(Context c) {
        mContext = c;
    }

    public int getCount() {
        return mThumbIds.length;
    }

    public Object getItem(int position) {
        return null;
    }

    public long getItemId(int position) {
        return 0;
    }

    public View getView(int position, View convertView, ViewGroup parent) {
        ImageView imageView;
        if (convertView == null) {
            imageView = new ImageView(mContext);
            int width = GridView.LayoutParams.MATCH_PARENT;
            int height = GridView.LayoutParams.MATCH_PARENT;
            imageView.setLayoutParams(new GridView.LayoutParams(width, height));
            imageView.setScaleType(ImageView.ScaleType.CENTER_INSIDE);
        } else {
            imageView = (ImageView) convertView;
        }

        imageView.setImageResource(mThumbIds[position]);
        return imageView;
    }

    private final Integer[] mThumbIds = {
            R.mipmap.png0001, R.mipmap.png0002,
```

```
            R.mipmap.png0003, R.mipmap.png0004,
            R.mipmap.png0007, R.mipmap.png0008,
            R.mipmap.png0009, R.mipmap.png0010,
            R.mipmap.png0001, R.mipmap.png0002,
            R.mipmap.png0003, R.mipmap.png0004,
            R.mipmap.png0005, R.mipmap.png0006,
            R.mipmap.png0009, R.mipmap.png0010,
            R.mipmap.png0001, R.mipmap.png0002,
            R.mipmap.png0003, R.mipmap.png0004,
            R.mipmap.png0005, R.mipmap.png0006,
            R.mipmap.png0007, R.mipmap.png0008
    };
}
```

在 ImageAdapter 类的 getView()方法中，convertView 是可重用的对象，它是在使用 GridView 组件显示数据时，Adapter 接口对象返回给 GridView 组件的 ImageView 对象，但是由于用户滚动屏幕，因此这个对象即将从屏幕上消失。Java 虚拟机在创建和销毁对象时是比较花费时间的，Android 将这个即将被销毁的对象返回给 getView()方法，在该方法中判断是否可以重用这个对象，以便提升应用程序的运行效率。在 getView()方法中，先判断 convertView 对象的值是否为空，如果不为空，则表示可以重用 convertView 对象，从而减少重新创建一个 ImageView 对象的时间。在 getView()方法中，还会对即将显示在 GridView 组件中的 ImageView 对象进行布局：通过代码指定 ImageView 组件在水平方向和垂直方向上均占满 GridView 组件为 ImageView 对象分配的显示空间。

修改 MainActivity.java 文件中的代码，在 onCreate()方法中获取界面中 GridView 组件的引用，监听对 GridView 组件中各选项的点击事件，并且使用 Toast 组件显示被点击的图片序号，修改后的代码如下：

```
package com.example.ch0512;

import androidx.appcompat.app.AppCompatActivity;

import android.os.Bundle;
import android.view.View;
import android.widget.AdapterView;
import android.widget.GridView;
import android.widget.Toast;

public class MainActivity extends AppCompatActivity {

    @Override
    protected void onCreate(Bundle savedInstanceState) {
        super.onCreate(savedInstanceState);
        setContentView(R.layout.activity_main);
```

```
        GridView gridview = (GridView) findViewById(R.id.gridview);
        gridview.setAdapter(new ImageAdapter(this));
        gridview.setOnItemClickListener(new AdapterView.OnItemClickListener() {
            @Override
            public void onItemClick(AdapterView<?> parent, View view, int position, long id) {
                Toast.makeText(MainActivity.this, "" + position, Toast.LENGTH_SHORT).
                    show();
            }
        });
    }
}
```

运行 Ch0512 应用程序，即可得到图 5-21 所示的运行效果。在点击某个图片后，会在 Toast 组件中显示被点击的图片序号。

5.7 同步练习三

1．完善 5.6.3 节中第 2 部分的案例，使其显示书的名称、封面图片、作者和出版社名称。
2．Spinner 组件探究：读者可以查询资料，通过自我探究修改 Spinner 组件在展开下拉列表时的显示样式。

5.8 Android 中的其他常用组件

在 Android 平台的 SDK 中，android.widget 包中还有许多组件，Android 中常用的组件及其含义如表 5-5 所示。

表 5-5　Android 中常用的组件及其含义

组 件 名 称	含 义
CalendarView	日历组件，用户可以从中选择日期
Chronometer	一个简单的计时器组件
DatePicker	日期选择组件
HorizontalScrollView	带有水平滚动条的布局容器组件
ImageButton	图片按钮组件
NumberPicker	数字选择器组件，允许用户在指定范围内选择一个整数
ProgressBar	进度条组件
Space	在布局中分隔组件，从而使组件之间留下间隙
TabHost	选项卡组件
TableLayout	表格布局容器组件
TableRow	与 TableLayout 容器组件配合使用的表格布局中的一行

续表

组 件 名 称	含　义
TextClock	文本时针组件，以指定格式显示当前的日期和时间
TimePicker	时间选择组件
ViewFlipper	翻转视图组件，以动画方式切换显示的组件
WebView	网页视图组件，用于显示 Web 页面的组件

注：这个列表不完整，完整的内容可以参考 Android 帮助文档。

5.9　同步练习四

自我学习：编写一个 Android 应用程序，使其可以用 WebView 组件浏览网页。对于 WebView 组件的使用方法，读者可以参考 Android 帮助文档。

第 6 章

样式和主题

在进行 Android 应用程序的界面设计时，经常需要对界面及界面中的组件设置统一的显示外观，如界面的背景颜色、文本字号、文本颜色、组件的显示大小、是否显示标题栏等。可以为每个组件设置自己的显示属性，但是为了便于对外观进行统一管理，需要将这些外观设置集中起来。Android 是通过样式（Style）完成这项工作的。要在 Android 中使用样式定制外观，需要完成两方面工作：定义样式，将定义好的样式应用于界面中。

6.1 样式入门

下面举例说明 Android 是如何定义样式及将定义好的样式应用于界面中的。在 Android Studio 中新建一个名为 Ch0601 的 Android 应用程序工程，然后修改布局文件 res/layout/activity_main.xml 中的代码，修改后的代码如下：

```xml
<?xml version="1.0" encoding="utf-8"?>
<LinearLayout
    xmlns:android="http://schemas.android.com/apk/res/android"
    android:layout_width="match_parent"
    android:layout_height="match_parent"
    android:orientation="vertical" >

    <android.widget.Button
        android:layout_width="match_parent"
        android:layout_height="0dp"
        android:layout_weight="1"
        android:text="@string/text_btn_01" />

    <android.widget.Button
        android:layout_width="match_parent"
        android:layout_height="0dp"
        android:layout_weight="1"
        android:text="@string/text_btn_02" />

    <TextView
        android:layout_width="match_parent"
```

```
            android:layout_height="0dp"
            android:layout_weight="1"
            android:gravity="center"
            android:text="@string/text_textview" />

</LinearLayout>
```

这个布局文件很简单，在一个 LinearLayout 容器组件中放置了 2 个 Button 组件和 1 个 TextView 组件，并且这 3 个组件平分 LinearLayout 容器组件的显示空间。

修改 res/values/strings.xml 文件中的代码，在其中定义布局文件中引用的字符串资源，修改后的代码如下：

```
<resources>
    <string name="app_name">Ch0601</string>

    <string name="text_btn_01">第一个按钮</string>
    <string name="text_btn_02">第二个按钮</string>
    <string name="text_textview">学好 Android 的样式和主题</string>

</resources>
```

运行 Ch0601 应用程序，运行效果如图 6-1 所示。

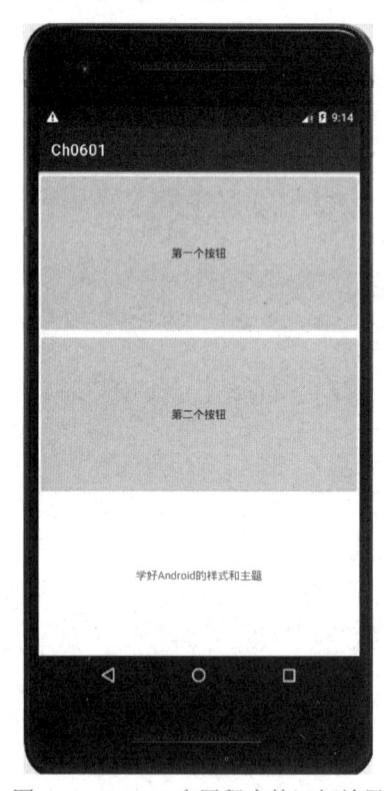

图 6-1 Ch0601 应用程序的运行效果

根据图 6-1 可知，2 个 Button 组件和 1 个 TextView 组件使用的都是 Android 自定义的默认样式，下面修改 Button 组件和 TextView 组件的样式。在工程的 res/values 目录下，新建一个名为 mystyles.xml 的文件，修改该文件中的代码，修改后的代码如下：

```xml
<resources>

    <style name="MyButtonStyle">
        <item name="android:textColor">#00FF00</item><!--绿色-->
        <item name="android:background">#FF0000</item><!--红色-->
        <item name="android:textSize">16sp</item>
    </style>

    <style name="MyTextViewStyle">
        <item name="android:textColor">#0000FF</item><!--黑色-->
        <item name="android:typeface">monospace</item>
    </style>

</resources>
```

在 res/values/mystyles.xml 文件中定义了两个新的样式，一个是名为 MyButtonStyle 的样式（简称 MyButtonStyle 样式），另一个是名为 MyTextViewStyle 的样式（简称 MyTextViewStyle 样式）。在 MyButtonStyle 样式中定义了文本的颜色、背景颜色和字号，在 MyTextViewStyle 样式中定义了文本的颜色和字体。将 MyButtonStyle 样式应用于界面中的第一个 Button 组件上，将 MyTextViewStyle 样式应用于界面中的 TextView 组件上。因此，修改 res/layout/activity_main.xml 文件中的代码，修改后的代码如下：

```xml
<?xml version="1.0" encoding="utf-8"?>
<LinearLayout
    xmlns:android="http://schemas.android.com/apk/res/android"
    android:layout_width="match_parent"
    android:layout_height="match_parent"
    android:orientation="vertical" >

    <android.widget.Button
        android:layout_width="match_parent"
        android:layout_height="0dp"
        android:layout_weight="1"
        style="@style/MyButtonStyle"
        android:text="@string/text_btn_01" />

    <android.widget.Button
        android:layout_width="match_parent"
        android:layout_height="0dp"
        android:layout_weight="1"
```

```xml
        android:text="@string/text_btn_02" />

    <TextView
        android:layout_width="match_parent"
        android:layout_height="0dp"
        android:layout_weight="1"
        android:gravity="center"
        style="@style/MyTextViewStyle"
        android:text="@string/text_textview" />

</LinearLayout>
```

在第一个 Button 组件的代码中，以下代码表示将 MyButtonStyle 样式应用于当前 Button 组件上。

```
style="@style/MyButtonStyle"    //将定义好的 style 应用于这个组件上
```

在 TextView 组件的代码中，以下代码表示将 MyTextViewStyle 样式应用于当前 TextView 组件上。

```
style="@style/MyTextViewStyle"    //将定义好的 style 应用于这个组件上
```

再次运行 Ch0601 应用程序，运行效果如图 6-2 所示。

图 6-2　添加样式后的 Ch0601 应用程序的运行效果

比较图 6-1 和图 6-2，可以看出在添加样式前、后，Ch0601 应用程序运行效果的不同之处。通过这个案例，我们对 Android 中样式的定义和使用方法有了一个初步的了解。下面详细介绍如何定义和使用样式。

6.2 定义样式

6.2.1 定义样式的一般格式

要定义一个样式，可以在工程的 res/values 目录下新建一个 XML 文件，也可以在现有的某个文件中（如 styles.xml 文件）直接添加要定义的样式。定义样式的一般格式如下：

```xml
<?xml version="1.0" encoding="utf-8"?>
<resources>

    <style name="自定义样式名称" parent="父样式名称">
        <item name="样式属性名称">属性值</item>
        …
    </style>

    <style name="自定义样式名称" parent="父样式名称">
        <item name="样式属性名称">属性值</item>
        …
    </style>

    …

</resources>
```

在 Java 文件中，以 "R.style.自定义样式名称" 的格式访问定义的样式；在 XML 文件中，以 "@style/自定义样式名称" 的格式访问定义的样式。需要注意的是，在定义样式时，parent 属性是可选属性，parent="父样式名称" 表示定义的样式是支持继承的，这种样式是级联样式。下面来看一个定义样式的案例，代码如下：

```xml
<?xml version="1.0" encoding="utf-8"?>
<resources>

    <style name="GreenText" parent="@android:style/TextAppearance">
        <item name="android:textColor">#00FF00</item>
    </style>

</resources>
```

在上述代码中，定义了一个名为 GreenText 的样式，它继承了 Android 平台已定义的名为 @android:style/TextAppearance 的样式，并且将其中的 android:textColor 属性值修改为#00FF00，即绿色。下面通过继承定义新的样式。例如，定义一个名为 GreenTextLarge 的样式，使其继承 GreenText 样式，代码如下：

```xml
<?xml version="1.0" encoding="utf-8"?>
<resources>
```

```xml
    <style name="GreenText" parent="@android:style/TextAppearance">
        <item name="android:textColor">#00FF00</item>
    </style>

    <style name="GreenTextLarge" parent="@style/GreenText">
        <item name="android:textSize">32sp</item>
    </style>

</resources>
```

因为 GreenText 样式是自定义样式,所以可以使用以下方式定义继承 GreenText 样式的样式。

```xml
<?xml version="1.0" encoding="utf-8"?>
<resources>

    <style name="GreenText" parent="@android:style/TextAppearance">
        <item name="android:textColor">#00FF00</item>
    </style>

    <style name="GreenTextLarge" parent="@style/GreenText">
        <item name="android:textSize">32sp</item>
    </style>

    <style name="GreenText.Small">     //可以使用这种方式定义继承自定义父样式的样式
        <item name="android:textSize">8sp</item>
    </style>

</resources>
```

这里定义了一个名为 GreenText.Small 的样式,这是一种特殊的样式,表示 GreenText.Small 样式是一个新的样式,并且它的父样式是 GreenText 样式。

6.2.2 样式定义中的可用属性

样式的应用目标不同,样式定义中的可用属性也有所不同。例如,定义一个针对 TextView 组件的样式和一个针对 Button 组件的样式,这两个样式定义中的可用属性是不同的。因此,应该针对样式的应用目标、参考组件的可用 XML 配置属性,确定样式定义中的可用属性。但有一种例外,如果将某个样式应用于某个组件上,而在该样式定义中包含该组件不支持的属性,那么该组件会自动忽略这个不支持的属性,不会影响其他支持的属性所起的作用。

如果读者对 Android 支持的完整属性列表感兴趣,则可以参考 Android 帮助文档中的 android.R.styleable 类,在该类中对每个 Android 组件都给出了其样式定义中的可用属性。例如,在 android.R.styleable 类中查看 ImageView 组件样式定义中的部分可用属性,如图 6-3 所示。

由于 ImageView 组件是 View 组件的子组件,因此 View 组件样式定义中的可用属性也是 ImageView 组件样式定义中的可用属性,在 android.R.styleable 类中,View 组件样式定义中的部分可用属性如图 6-4 所示。

ImageView

`public static final int[] ImageView`

Attributes that can be used with a ImageView.
Includes the following attributes:

Attribute	Description
android:src	Sets a drawable as the content of this ImageView.
android:scaleType	Controls how the image should be resized or moved to match the size of this ImageView.
android:adjustViewBounds	Set this to true if you want the ImageView to adjust its bounds to preserve the aspect ratio of its drawable.
android:maxWidth	An optional argument to supply a maximum width for this view.

图 6-3 ImageView 组件样式定义中的部分可用属性

View

`public static final int[] View`

Attributes that can be used with View or any of its subclasses. Also see ViewGroup_Layout for attributes that are processed by the view's parent.
Includes the following attributes:

Attribute	Description
android:theme	The overall theme to use for an activity.
android:scrollbarSize	Sets the width of vertical scrollbars and height of horizontal scrollbars.
android:scrollbarThumbHorizontal	Defines the horizontal scrollbar thumb drawable.
android:scrollbarThumbVertical	Defines the vertical scrollbar thumb drawable.

图 6-4 View 组件样式定义中的部分可用属性

6.3 应用样式

在完成样式定义后，可以将样式应用于某个组件上，也可以将样式应用于某个 Activity 或整个 Application 上。

6.3.1 将样式应用于某个组件上

将定义好的样式应用于某个组件上是一项非常简单的工作，在组件的配置中添加 styleXML 配置属性即可。例如，要将前面定义的 GreenText.Small 样式应用于 TextView 组件上，只需添加 style 属性，代码如下：

```
<TextView
    style="@style/GreenText.Small"
```

```xml
...
android:text="@string/hello" />
```

可以将样式应用于具体组件上,也可以将样式应用于容器组件上。需要注意的是,应用于容器组件上的样式只对容器组件本身有效,对放置于该容器组件中的子组件是不起作用的。

6.3.2 将样式应用于某个 Activity 或整个 Application 上

首先介绍什么是主题。在将样式应用于某个 Activity 或整个 Application 上时,这个样式就成了主题。将样式应用于某个 Activity 或整个 Application 上,需要在 AndroidManifest.xml 文件中针对某个 Activity 或整个 Application 添加 android:theme 属性。

下面来看一个案例。首先定义一个样式,代码如下:

```xml
<?xml version="1.0" encoding="utf-8"?>
<resources>
    <color name="custom_theme_color">#b0b0ff</color>
    <style name="CustomTheme" parent="@style/MyTheme.Light">
        <item name="android:windowBackground">@color/custom_theme_color</item>
        <item name="android:colorBackground">@color/custom_theme_color</item>
    </style>
</ resources >
```

然后将这个样式应用于某个 Activity 上,代码如下:

```xml
<activity android:theme="@style/CustomTheme">
```

或者将这个样式应用于整个 Application 上,代码如下:

```xml
<application android:theme="@style/CustomTheme">
```

那么这个样式就是主题。

主题是一种特殊的样式,由于主题是应用于某个 Activity 或整个 Application 上的,因此 Android 为主题的定义引入了一些特殊的属性。例如,android:windowNoTitle 属性只能应用于对主题的定义中,该属性主要用于在显示某个 Activity 或整个 Application 时,设置是否显示 Activity 或 Application 的标题。对于主题定义中的可用属性,可以参考 Android 帮助文档的 android.R.styleable 类中 Theme 的相关内容。主题定义中的部分可用属性如图 6-5 所示。

Theme

public static final int[] Theme

These are the standard attributes that make up a complete theme.
Includes the following attributes:

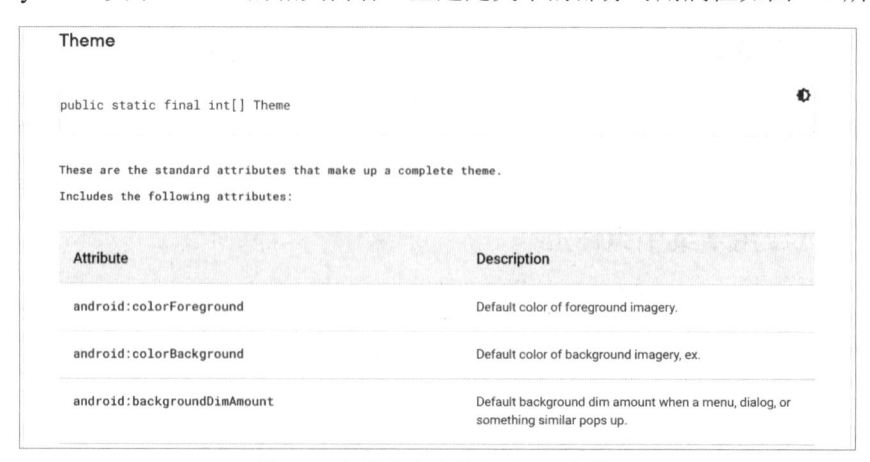

Attribute	Description
android:colorForeground	Default color of foreground imagery.
android:colorBackground	Default color of background imagery, ex.
android:backgroundDimAmount	Default background dim amount when a menu, dialog, or something similar pops up.

图 6-5 主题定义中的部分可用属性

6.4 使用 Android 平台已定义的样式和主题

Android 平台预定义了一系列的样式和主题，以供应用程序使用。在定义的所有样式中，以 Theme 开头的样式是主题，不以 Theme 开头的样式是普通样式。对于 Android 中完整的样式定义，读者可以参考 Android 帮助文档中的 android.R.style 类。要使用 Android 平台已定义的样式或主题，需要将样式或主题名中的下画线"_"替换为小数点"."。例如，要在应用程序的某个 Activity 中使用 Theme_NoTitleBar 主题，代码如下：

```
<activity android:theme="@android:style/Theme.NoTitleBar">
```

Android 平台已定义的部分典型的样式和主题如表 6-1 所示。

表 6-1 Android 平台已定义的部分典型的样式和主题

名　　称	类　　型	描　　述
Animation	样式	Android 动画的基础样式
DeviceDefault_ButtonBar	样式	Android 工具条的基础样式
Holo_ButtonBar	样式	基于 Holo 风格的工具条样式
MediaButton_(XXX)	样式	媒体播放器的样式
TextAppearence	样式	文本显示的基础样式
TextAppearence_(XXX)	样式	文本显示的变体样式
Theme	主题	Android 的基础主题
Theme_Black	主题	Android 的黑色主题
Theme_Black_NoTitleBar	主题	Android 的黑色主题，但不显示 Activity 的标题栏
Theme_Dialog	主题	在将 Activity 显示为对话框时使用的主题
Theme_Holo	主题	Holo 风格的基础主题
Theme_Holo_Dialog	主题	Holo 风格的对话框主题
Theme_Holo_Light	主题	Holo 风格的明亮主题
Theme_Holo_Light_NoActionBar	主题	Holo 风格的明亮主题，但不显示 Activity 的导航栏
Theme_Light	主题	显示明亮背景和黑色字体的主题
Theme_Light_NoTitleBar_FullScreen	主题	全屏显示明亮背景和黑色字体的主题，不显示 Activity 的标题栏
Theme_Material	主题	Material 风格的基础主题
Theme_Material_Light	主题	Material 风格的明亮主题

6.5 Android 应用程序的主题结构分析

在介绍完样式与主题的相关知识后，下面介绍 Android 应用程序中与主题有关的内容。

在 Android Studio 中新建 Android 应用程序工程时，Android 已经为该 Android 应用程序设置了默认的主题。打开 AndroidManifest.xml 文件，代码如下：

```
<?xml version="1.0" encoding="utf-8"?>
<manifest xmlns:android="http://schemas.android.com/apk/res/android"
    package="com.example.ch0601">
```

```xml
<application
    android:allowBackup="true"
    android:icon="@mipmap/ic_launcher"
    android:label="@string/app_name"
    android:roundIcon="@mipmap/ic_launcher_round"
    android:supportsRtl="true"
    android:theme="@style/Theme.Ch0601">
    <activity
        android:name=".MainActivity"
        android:exported="true">
        <intent-filter>
            <action android:name="android.intent.action.MAIN" />

            <category android:name="android.intent.category.LAUNCHER" />
        </intent-filter>
    </activity>
</application>

</manifest>
```

以下代码指定了该应用程序的主题为 Theme.Ch0601。

```
android:theme="@style/Theme.Ch0601" >        //用于指定该应用程序的主题
```

在 res/values 目录下的 themes.xml 文件中，可以找到对 Theme.Ch0601 主题的定义。res/values/themes.xml 文件中的代码如下：

```xml
<resources xmlns:tools="http://schemas.android.com/tools">
    <!-- Base application theme. -->
    <style name="Theme.Ch0601"
        parent="Theme.MaterialComponents.DayNight.DarkActionBar">
        <!-- Primary brand color. -->
        <item name="colorPrimary">@color/purple_500</item>
        <item name="colorPrimaryVariant">@color/purple_700</item>
        <item name="colorOnPrimary">@color/white</item>
        <!-- Secondary brand color. -->
        <item name="colorSecondary">@color/teal_200</item>
        <item name="colorSecondaryVariant">@color/teal_700</item>
        <item name="colorOnSecondary">@color/black</item>
        <!-- Status bar color. -->
        <itemname="android:statusBarColor"
            tools:targetApi="l">?attr/colorPrimaryVariant</item>
        <!-- Customize your theme here. -->
    </style>
</resources>
```

6.6 同步练习

Android 平台上预定义了很多样式和主题，将表 6-1 中的样式和主题应用于一个 Android 应用程序中，观察每个样式或主题的外观。

第 7 章

理解和使用 Intent

在基于 HTML 的页面应用程序中，我们使用超链接实现页面之间的跳转。前面提到过，Android 应用程序界面是由一个或多个 Activity 组成的，一个 Activity 相当于 HTML 中的一个页面，那么，当一个 Android 应用程序具有多个相互联系的 Activity 时，它们之间是如何实现跳转的呢？它们是通过 Intent 实现跳转的。

Intent 不仅可以实现 Activity 之间的跳转，还是 Android 平台上各个部分之间实现信息沟通的桥梁。本章将对 Intent 的相关知识进行详细的介绍。

7.1　Intent 入门

下面通过一个简单的案例说明什么是 Intent，以及 Intent 的基本应用。

这个案例应用程序的目标是，首先显示一个 Activity，在该 Activity 中使用 TextView 组件显示一张图片的名称，使用 Button 组件表示一个按钮，点击这个按钮，即可在一个新的 Activity 中显示这张图片。非常简单，不是吗？但它能很好地帮助读者理解 Intent。

在 Android Studio 中新建一个名为 Ch0701 的 Android 应用程序工程，然后修改主界面布局文件 res/layout/activity_main.xml 中的代码，修改后的代码如下：

```xml
<?xml version="1.0" encoding="utf-8"?>
<LinearLayout xmlns:android="http://schemas.android.com/apk/res/android"
    android:layout_width="match_parent"
    android:layout_height="match_parent"
    android:orientation="vertical">

    <TextView
        android:layout_width="match_parent"
        android:layout_height="0dp"
        android:layout_weight="1"
        style="@android:style/TextAppearance.Holo.Large"
        android:gravity="center"
        android:text="@string/text_beauty" />

    <View
        android:layout_width="match_parent"
        android:layout_height="10dp"
```

```xml
        android:background="#000"/>

    <Button
        android:id="@+id/id_button"
        android:layout_width="match_parent"
        android:layout_height="0dp"
        android:layout_weight="1"
        android:text="@string/text_button" />

</LinearLayout>
```

这个布局文件中包含一个 TextView 组件和一个 Button 组件，两个组件之间的间隔为 10dp。此外，对 TextView 组件使用了 Android 已定义的文本样式，使显示的文本更大。

为了能在一个新的 Activity 中显示一张图片，我们需要为这个 Activity 创建一个布局文件。因此，在 res/layout 目录下，新建一个名为 layout02.xml 的布局文件，该文件中的代码如下：

```xml
<?xml version="1.0" encoding="utf-8"?>
<LinearLayout xmlns:android="http://schemas.android.com/apk/res/android"
    android:layout_width="match_parent"
    android:layout_height="match_parent"
    android:orientation="vertical" >

    <ImageView
        android:layout_width="match_parent"
        android:layout_height="match_parent"
        android:src="@drawable/beauty"
        android:scaleType="center"
        android:contentDescription="@string/text_empty" />

</LinearLayout>
```

我们需要将一张图片存储于 res/mipmap 目录下。由于布局文件中引用了字符串资源，因此需要修改 res/values/strings.xml 文件中的代码，修改后的代码如下：

```xml
<?xml version="1.0" encoding="utf-8"?>
<resources>

    <string name="app_name">Ch0701</string>

    <string name="text_beauty">一张美丽的风景照片</string>
    <string name="text_empty">""</string>
    <string name="text_button">点击按钮显示最美风景</string>

</resources>
```

如果此时运行 Ch0701 应用程序，那么可以正确显示第一个 Activity，但是当点击按钮时，不会显示第二个 Activity。

为了显示第二个 Activity，我们需要创建一个新的 Activity，使之显示 layout02.xml 文件的界面。为此，在 Ch0701 应用程序工程的 src 目录下的 com.example.ch0701 包中，新建一个名为 Activity02 的 Java 类文件。Activity02.java 文件中的代码如下：

```java
package com.ttt.ex07intent01;

import android.app.Activity;
import android.os.Bundle;

public class Activity02 extends Activity {

    @Override
    protected void onCreate(Bundle savedInstanceState) {
        super.onCreate(savedInstanceState);
        setContentView(R.layout.layout02);
    }
}
```

现在有两个 Activity，分别为 MainActivity 和 Activity02，它们分别显示 activity_main.xml 文件和 layout02.xml 文件的界面。可是如何将它们关联起来呢？也就是如何在点击第一个界面中的按钮时，启动第二个 Activity，从而显示 layout02.xml 文件的界面呢？为此，需要修改 MainActivity.java 文件中的代码，使其监听对按钮的点击事件，并且对点击事件进行相应的响应，修改后的代码如下：

```java
package com.ttt.ex07intent01;

import android.content.Intent;
import android.os.Bundle;
import android.support.v7.app.AppCompatActivity;
import android.view.View;
import android.widget.Button;

public class MainActivity extends AppCompatActivity
                          implements View.OnClickListener {

    @Override
    protected void onCreate(Bundle savedInstanceState) {
        super.onCreate(savedInstanceState);
        setContentView(R.layout.activity_main);

        Button btn = (Button)this.findViewById(R.id.id_button);
        btn.setOnClickListener(this);
    }

    @Override
```

```
    public void onClick(View v) {
        Intent i = new Intent(this, Activity02.class);
        this.startActivity(i);
    }
}
```

在对按钮点击事件的响应方法 onClick() 中，先创建一个 Intent 对象 i，再使用 startActivity(i) 方法启动指定的 Activity。在运行 Ch0701 应用程序前，还需要在 AndroidManifest.xml 文件中登记 Activity02。因为 Android 规定，所有的 Activity 都必须在工程的 AndroidManifest.xml 文件中登记。因此，修改 AndroidManifest.xml 文件中的代码，修改后的代码如下：

```xml
<?xml version="1.0" encoding="utf-8"?>
<manifest xmlns:android="http://schemas.android.com/apk/res/android"
    package="com.example.ch0701">

    <application
        android:allowBackup="true"
        android:icon="@mipmap/ic_launcher"
        android:label="@string/app_name"
        android:roundIcon="@mipmap/ic_launcher_round"
        android:supportsRtl="true"
        android:theme="@style/Theme.Ch0701">
        <activity
            android:name=".MainActivity"
            android:exported="true">
            <intent-filter>
                <action android:name="android.intent.action.MAIN" />
                <category android:name="android.intent.category.LAUNCHER" />
            </intent-filter>
        </activity>

        <activity
            android:name=".Activity02"
            android:exported="false" >
        </activity>

    </application>

</manifest>
```

在 AndroidManifest.xml 文件的 <application> 标签中，通过 <activity> 标签登记 Activity02。现在运行 Ch0701 应用程序，运行效果如图 7-1 所示。点击按钮，会在一个新的界面中显示一张美丽的风景照片，如图 7-2 所示。这正是我们期望的效果。

图 7-1 Ch0701 应用程序的运行效果   图 7-2 点击按钮后的界面

7.2 同步练习一

编写一个与 7.1 节中的案例相似、能通过 Intent 对象打开新的 Activity 的 Android 应用程序。例如，读者可以在点击一张图片后，显示一个文本框，用于显示该图片的基本信息。

7.3 细说 Intent

Android 的 Intent 对象是联系各个 Activity 的关键对象。Intent 的中文含义是意图，我们可以这样理解 Intent：通过 Intent 对象告诉 Android 要做什么。例如，在 7.1 节的案例中，我们创建了一个 Intent 对象，代码如下：

```
Intent i = new Intent(this, Activity02.class);
this.startActivity(i);
```

上述代码表示我们希望 Android 平台启动 Activity02。按照这种方式创建的 Intent 称为显式 Intent：在 Intent 中，我们明确地告诉 Android 平台要启动的 Activity。还有一种 Intent 称为隐式 Intent：在 Intent 对象中指定一些条件，Android 平台会根据这些条件启动最符合条件的 Activity。

为了便于读者理解隐式 Intent，下面对 7.1 节中的 Ch0701 应用程序工程进行修改。首先，修改 AndroidManifest.xml 文件中的代码，修改后的代码如下：

```xml
<?xml version="1.0" encoding="utf-8"?>
<manifest xmlns:android="http://schemas.android.com/apk/res/android"
    package="com.example.ch0701">

    <application
        android:allowBackup="true"
        android:icon="@mipmap/ic_launcher"
        android:label="@string/app_name"
        android:roundIcon="@mipmap/ic_launcher_round"
        android:supportsRtl="true"
        android:theme="@style/Theme.Ch0701">
        <activity
            android:name=".MainActivity"
            android:exported="true">
            <intent-filter>
                <action android:name="android.intent.action.MAIN" />
                <category android:name="android.intent.category.LAUNCHER" />
            </intent-filter>
        </activity>

        <activity
            android:name=".Activity02"
            android:exported="false">
            <intent-filter>
                <action android:name="com.example.ch0701.A1" />
                <category android:name="android.intent.category.DEFAULT" />
            </intent-filter>
        </activity>

    </application>

</manifest>
```

注意以下代码。

```xml
            <intent-filter>
                <action android:name="com.example.ch0701.A1" />
                <category android:name="android.intent.category.DEFAULT" />
            </intent-filter>
```

上述代码为Activity02定义了一个Intent过滤器，表示在创建Intent对象时，只要指定Intent对象的 action 为 com.example.ch0701.A1，并且指定 Intent 对象的 category 为 android.intent.category.DEFAULT，即可启动Activity02。此外，Android 允许为一个 Activity 定义多个 Intent 过滤器。为了能够采用隐式的方式启动Activity02，修改 MainActivity.java 文件中的代码，修改后的代码如下：

```java
package com.example.ch0701;

import androidx.appcompat.app.AppCompatActivity;

import android.content.Intent;
import android.os.Bundle;
import android.view.View;
import android.widget.Button;

public class MainActivity extends AppCompatActivity
                                    implements View.OnClickListener {

    @Override
    protected void onCreate(Bundle savedInstanceState) {
        super.onCreate(savedInstanceState);
        setContentView(R.layout.activity_main);

        Button btn = (Button)this.findViewById(R.id.id_button);
        btn.setOnClickListener(this);
    }

    @Override
    public void onClick(View v) {
        Intent i = new Intent("com.example.ch0701.A1");
        i.addCategory(Intent.CATEGORY_DEFAULT);
        this.startActivity(i);
    }

}
```

注意创建 Intent 对象和启动 Activity02 的代码。

```java
Intent i = new Intent("com.example.ch0701.A1");
i.addCategory(Intent.CATEGORY_DEFAULT);
this.startActivity(i);
```

在上述代码中，在创建 Intent 对象时指定了要启动的 Activity 支持的 action，并且使用 addCategory()方法指定了当前动作（action）被执行的环境，因此，当执行代码 this.startActivity(i) 时，Android 平台会启动满足这些条件的 Activity，本案例中是 Activity02。这里有一个问题：当 Intent 对象中有多个 Activity 满足指定的条件时，Android 平台会怎样处理呢？Android 平台会弹出一个对话框，让用户选择启动哪个满足条件的 Activity。

运行修改后的 Ch0701 应用程序，即可得到与图 7-1 和图 7-2 一致的运行效果。

根据该案例可知 AndroidManifest.xml 文件是如何对 Activity 进行配置的。在<activity>标签中，<intent-filter>标签所指定的属性与创建 Intent 对象时所指定的属性是密切相关的，它们必须匹配才能启动我们需要的 Activity。隐式 Intent 的配置过程如图 7-3 所示。

```
某个Activity ──1、创建──▶  Intent i = new Intent("com.example.ch0701.A1");
                          i.addCategory(Intent.CATEGORY_DEFAULT);
                          this.startActivity(i);

                                    │
                                  2、匹配
                                    ▼

Activity02  ◀──3、启动──  <activity
                              android:name=".Activity02"
                              android:exported="false">
                              <intent-filter>
                                  <action android:name="com.example.ch0701.A1" />
                                  <category android:name="android.intent.category.DEFAULT" />
                              </intent-filter>
                          </activity>
```

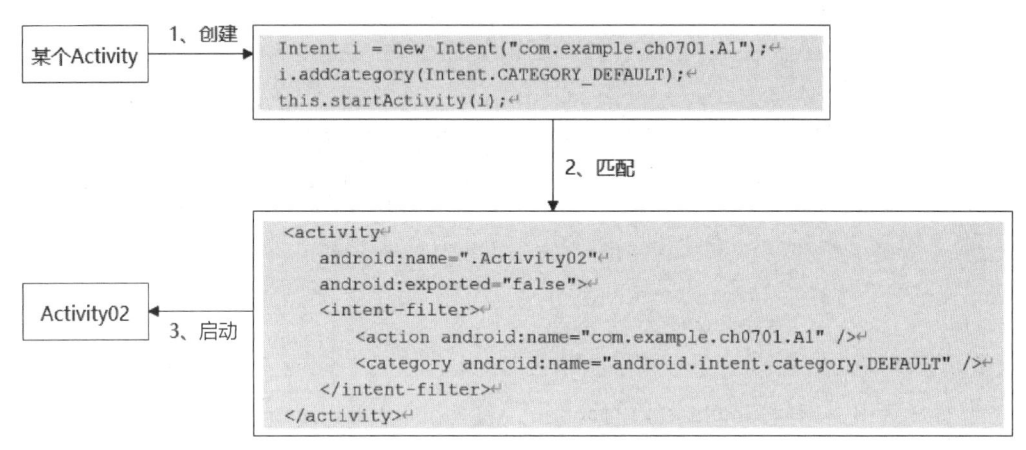

图 7-3　隐式 Intent 的配置过程

根据图 7-3 可知，隐式 Intent 的配置过程如下：首先，某个 Activity 在 AndroidManifest.xml 文件中配置自己，包括配置自己的<intent-filter>标签；然后，另一个 Activity 创建一个 Intent 对象，创建的 Intent 对象中指定的 action 和 category 与之前的 Activity 在其<intent-filter>标签中设置的 action 和 category 相匹配，所以，当调用 startActivity(i)方法时，会启动该 Intent 对象指定的 Activity。

那么，在创建 Intent 对象时，可以指定哪些条件呢？或者说，在 Activity 的<intent-filter>标签中，可以设置哪些条件呢？这些条件包括 action、data 和 category。

7.3.1　Intent 的 action

Intent 的 action 主要用于启动 Activity。action 是一个字符串常量，我们可以任意定义，但是，Android 建议采用"Java 包名+特定字符串"的形式为 action 命名。例如，将 Activity02 的 action 命名为 com.example.ch0701.A1，其中 com.example.ch0701 是包名，A1 是特定的名称。需要注意的是，Activity 与 action 不一定是一对一的关系，也就是说，可能多个 Activity 对应同一个 action。当出现这种情况时，Android 平台会弹出一个对话框，用于选择满足条件的 Activity。

Intent 中预定义了一些常用的 action，包括 Intent.ACTION_MAIN、Intent.ACTION_VIEW、Intent.ACTION_EDIT、Intent.ACTION_PICK、Intent.ACTION_DIAL、Intent.ACTION_CALL、Intent.ACTION_DELETE、Intent.ACTION_INSERT、Intent.ACTION_SEARCH 等，每个 action 都有明确的含义。例如，Intent.ACTION_MAIN 表示包含这个 action 的 Activity 是 android 应用程序的入口 Activity，也就是说，当运行某个 Android 应用程序时，首先打开包含这个 action 的 Activity；Intent.ACTION_PICK 表示从某个列表中选择一条信息。

使用 Intent 中预定义的 action 可以打开 Android 平台上的 Activity，如电话拨号 Activity。在 Android 应用程序中，可以直接使用这些 action 作为 Activity 的 action。为 Intent 对象指定 action 的方式有以下两种。

- 通过 Intent 类的构造方法 new Intent(String action)及 new Intent(String action, URI uri)。
- 通过 Intent 类的 setAction(String action)方法。

要指定一个 Activity 可以被哪些 action 打开，需要在 AndroidManifest.xml 文件中使用<intent-filter>标签说明，代码如下：

```xml
<activity class=".NotesList" android:label="@string/title_notes_list">
    <intent-filter>
        <action android:name="android.intent.action.VIEW" />
        <action android:name="android.intent.action.EDIT" />
        <action android:name="android.intent.action.PICK" />
        <category android:name="android.intent.category.DEFAULT" />
        <data android:mimeType="vnd.android.cursor.dir/vnd.google.note" />
    </intent-filter>
</activity>
```

在当前案例中，只有将 Intent 对象的 action 设置为 android.intent.action.VIEW、android.intent.action.EDIT 和 android.intent.action.PICK 中的一个时，才可以打开 NotesList 中的 Activity。需要注意的是，在 AndroidManifest.xml 文件中设置 action 的 android:name 属性时，使用的是常量名称对应的字符串，而不是常量名称。此外，要打开这个 Activity，需要将 Intent 的 category 设置为 android.intent.category.DEFAULT，以及将 Intent 的 data 的 mimeType 属性值设置为 vnd.android.cursor.dir/vnd.google.note。

7.3.2 Intent 的 data

在使用隐式 Intent 打开 Activity 时，除了指定 Activity 的 action，通常还会指定 Activity 所支持的 data。就像在 HTML 页面中那样，除了指明 GET、PUT、POST 操作，通常还需要指定页面的 URI 地址。在 Android 应用程序中,通过 Intent 对象的 data 指定要操作的数据。Android 中的 data 是通过 URI 指定的。URI 的标准形式为 scheme://host:port/path、scheme://host:port/pathPattern 或 scheme://host:port/pathPrefix。例如，在 URI content://com.example.project:200/folder/subfolder/etc 中，scheme 为 "content"，host 为 "com.example.project"，port 为 "200"，path 为 "folder/subfolder/etc"。

Activity 的 data 是在 AndroidManifest.xml 文件中 Activity 配置的<intent-filter>标签中通过<data>标签指定的，data 中的属性（简称 data 属性）包括 host、mimeType、path、pathPattern、pathPrefix、port 和 scheme。指定 data 属性的一般格式如下：

```xml
<data android:host="string"
    android:mimeType="string"
    android:path="string"
    android:pathPattern="string"
    android:pathPrefix="string"
    android:port="string"
    android:scheme="string" />
```

如果要使用 Intent 对象启动设置了 data 属性的某个 Activity，则需要在创建 Intent 对象时，明确设置 data 属性。在 Intent 中设置 data 属性的方法如下。

- setData(Uri data)。设置 Intent 对象的 data 属性，该方法会自动清除之前设置的 mimeType 属性。

- setDataAndType(Uri data, String mimeType)。同时设置 Intent 对象的 data 属性和 mimeType 属性。
- setType(String mimeType)。设置 Intent 对象的 mimeType 属性,该方法会自动清除之前设置的 data 属性。

下面举例说明如何在<intent-filter>标签中设置 data 属性,并且使用隐式 Intent 启动设置了 data 属性的 Activity。

在 AndroidManifest.xml 文件中配置某个 Activity,代码如下:

```xml
<activity class=".NotesList" android:label="@string/title_notes_list">
    <intent-filter>
        <action android:name="android.intent.action.MAIN" />
        <category android:name="android.intent.category.LAUNCHER" />
    </intent-filter>
    <intent-filter>
        <action android:name="android.intent.action.VIEW" />
        <action android:name="android.intent.action.EDIT" />
        <action android:name="android.intent.action.PICK" />
        <category android:name="android.intent.category.DEFAULT" />
        <data android:scheme="content" />
        <data android:mimeType="vnd.android.cursor.dir/vnd.google.note" />
    </intent-filter>
</activity>
```

在第一个<intent-filter>标签中,将 action 设置为 android.intent.action.MAIN,表示该 Activity 是相应 Android 应用程序的主入口 Activity;将 category 设置为 android.intent.category.LAUNCHER,表示该 Activity 会出现在 Android 的 HOME 界面的应用程序列表中。第二个<intent-filter>标签中配置了 3 个 action,分别为 android.intent.action.VIEW、android.intent.action.EDIT 和 android.intent.action.PICK;配置了 1 个 category,即 android.intent.category.DEFAULT;设置 data 的 scheme 属性值为 content、mimeType 属性值为 vnd.android.cursor.dir/vnd.google.note。创建一个 Intent 对象,然后采用隐式方式启动该 Activity,代码如下:

```
Uri uri = Uri.parse("content://com.example.project:200/folder/subfolder/etc");
Intent i = new Intent("android.intent.action.EDIT");
i.setDataAndType(uri, "vnd.android.cursor.dir/vnd.google.note");
this.startActivity(i);
```

我们已经在 AndroidManifest.xml 文件中设置了 Activity 的 category,但是在创建 Intent 对象时,并没有设置 category,可以打开将 category 设置为 DEFAULT 的 Activity 吗?答案是可以。因为 Android 在创建 Intent 对象时,会自动将 Intent 对象的 category 设置为 DEFAULT。

7.3.3 Intent 的 category

category 的含义是类别,它是 Android 对 Activity 进行分类的一种手段。例如,如果将某个 Activity 的 category 设置为 LAUNCHER,那么 Android 会在启动时将该 Activity 显示在 Home

界面中，以便运行该 Activity；如果将某个 Activity 的 category 设置为 HOME，那么 Android 会在启动时将该 Activity 作为 Android 平台的 HOME 界面。

Intent 中预定义了一些 category，常用的 category 包括 Intent.CATEGORY_DEFAULT、Intent.CATEGORY_LAUNCHER、Intent.CATEGORY_INFO、Intent.CATEGORY_HOME、Intent.CATEGORY_PREFERENCE、Intent.CATEGORY_CAR_DOCK、Intent.CATEGORY_CAR_MODE、Intent.CATEGORY_APP_MARKET 等。

下面来看一个案例，获取所有出现在 HOME 界面中的 Activity，代码如下：

```
Intent mainIntent = new Intent(Intent.ACTION_MAIN, null);
mainIntent.addCategory(Intent.CATEGORY_LAUNCHER);
PackageManager pm = getPackageManager();
List<ResolveInfo> list = pm.queryIntentActivities(mainIntent, 0);
for(ResolveInfo ri: list)
{
    Log.d("test",ri.toString());
    String packagename = ri.activityInfo.packageName;
    String classname = ri.activityInfo.name;
    Log.d("test", packagename + ":" + classname);
    if(classname.equals("com.ttt.ex07intent01.Activity02"))
    {
        Intent ni = new Intent();
        ni.setClassName(packagename,classname);
        activity.startActivity(ni);
    }
}
```

该案例应用程序的编写思路是，创建一个 action 为 Intent.ACTION_MAIN、category 为 Intent.CATEGORY_LAUNCHER 的 Intent 对象，然后通过 PackageManager 查询所有匹配该 Intent 对象的 Activity，并且显示匹配 Activity 的类名，如果一个 Activity 的类名等于指定的类名 "com.ttt.ex07intent01.Activity02"，则启动并运行该 Activity。

下面再看一个案例，通过以下代码，直接返回 Android 应用程序的 HOME 界面。

```
Intent mainIntent = new Intent(Intent.ACTION_MAIN, null);
mainIntent.addCategory(Intent.CATEGORY_HOME);
startActivity(mainIntent);
```

7.3.4 Intent 的 extra

在使用 Intent 对象启动并运行某个 Activity 时，有时需要将一些附加数据传递到被启动的 Activity 中，此时会用到 extra。extra 只作为传递给目标 Activity 的附加数据，不作为挑选 Activity 的匹配依据。

extra 是以 key/value 形式表示的数据，其中的 key 是 String 类型的键，value 可以是 Java 基本数据类型，也可以是实现了 android.os.Parcelable 接口的对象数据类型。Intent 提供了写入及读取基本数据类型数据的方法。例如，如果希望传递一个整数和一个字符串到目标 Activity 中，则可以在创建的 Intent 对象 intent 上执行以下代码。

```
intent.putExtra("productName", "iPhone");
intent.putExtra("ProductAmount", 100);
```

在目标 Activity 中，通过获取启动该 Activity 的 Intent 对象，可以获取传递过来的数据，代码如下：

```
String pn = intent.getStringExtra("productName");
int pa = intent.getIntExtra("ProductAmount");
```

7.4 Intent 解析

根据前面介绍的知识可知，Intent 对象与<intent-filter>标签是密切相关的：在使用 Intent 对象启动某个 Activity 时，必须将 Intent 对象中设置的属性（action、data 和 category）与<intent-filter>标签中配置的属性进行匹配，从而启动能够匹配的 Activity。我们将这种匹配操作称为 Intent 解析。在 Android 应用程序中，Intent 解析的步骤如下。

（1）在显式匹配中，在 Intent 对象中明确指定要启动的 Activity 的名称，方法包括 new Intent (Context packageContext, Class<?> cls)、setComponent(ComponentName)、setClass(Context, Class)。

（2）在隐式匹配中，分为以下几种情况。

- 如果在 Intent 对象中设置了 action 且设置的 action 出现在<intent-filter>标签中，或者在<intent-filter>标签中未设置任何 action，则匹配；如果在 Intent 对象中未设置 action，则匹配所有的 Activity。
- 如果在 Intent 对象中设置了 data 且设置的 data 出现在<intent-filter>标签中，则匹配；如果在 Intent 对象中未设置 data，则只能匹配未设置 data 的 Activity。
- 如果在 Intent 对象中设置了 category 且设置的 category 出现在<intent-filter>标签中，则匹配；如果在 Intent 对象中未设置 category，则只能匹配未设置 category 的 Activity。

7.5 获取 Activity 返回的结果

在前面的案例中，在使用 Intent 对象启动某个 Activity 时，并不需要被启动的目标 Activity 返回结果，而有时我们确实需要获取被启动的 Activity 返回的结果，该怎么做呢？

获取被启动 Activity 返回结果的方法不是 startActivity()，而是 startActivityForResult()。在使用 startActivityForResult()方法启动 Activity 时，在被启动的 Activity（称为目标 Activity）执行完毕后，Android 平台会调用源 Activity 的 onActivityResult()方法。

startActivityForResult()方法的原型如下：

```
public void startActivityForResult(Intent intent, int requestCode)
```

其中的 requestCode 是源 Activity 传递的一个识别参数。

在目标 Activity 执行完毕后，Android 平台会调用 onActivityResult()方法。onActivityResult() 方法的原型如下：

```
protected void onActivityResult(int requestCode, int resultCode, Intent data)
```

其中的 requestCode 是调用 startActivityForResult()方法时传递的 requestCode 参数，resultCode 是执行目标 Activity 的返回结果，其值可以为 Activity.RESULT_OK、Activity.RESULT_CANCELED

或其他自定义的值。

要获取目标 Activity 返回结果的前提是目标 Activity 有返回结果。目标 Activity 使用 setResult()方法设置返回结果。setResult()方法的原型如下：

```
public final void setResult(int resultCode)
public final void setResult(int resultCode, Intent data)
```

也就是说，目标 Activity 也是通过 Intent 对象给源 Activity 返回结果的。

下面举例说明如何启动一个 Activity，并且接收从目标 Activity 返回的结果。该案例应用程序运行效果的首界面如图 7-4 所示。点击按钮，界面中会显示两张图片，如图 7-5 所示。此时，点击任意一张图片，都可以关闭该 Activity，并且在首界面中显示所点击图片的编号，如图 7-6 所示。

图 7-4　案例应用程序运行效果的首界面 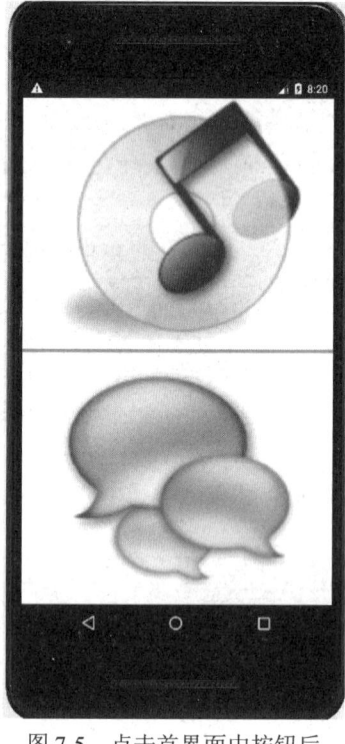　　图 7-5　点击首界面中按钮后显示的界面 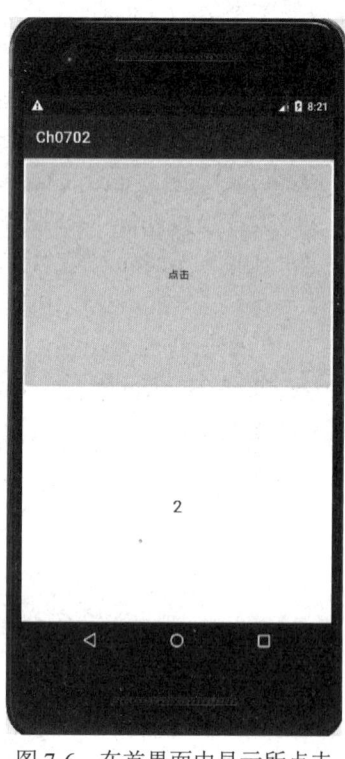　　图 7-6　在首界面中显示所点击图片的编号

下面构建该案例应用程序。在 Android Studio 中新建一个名为 Ch0702 的 Android 应用程序工程。修改主界面布局文件 res/layout/activity_main.xml 中的代码，修改后的代码如下：

```xml
<?xml version="1.0" encoding="utf-8"?>
<LinearLayout xmlns:android="http://schemas.android.com/apk/res/android"
    android:layout_width="match_parent"
    android:layout_height="match_parent"
    android:orientation="vertical">

    <Button
        android:id="@+id/id_button"
```

```xml
        android:layout_width="match_parent"
        android:layout_height="0dp"
        android:layout_weight="1"
        android:text="@string/text_button" />

    <TextView
        android:id="@+id/id_textview"
        android:layout_width="match_parent"
        android:layout_height="0dp"
        android:layout_weight="1"
        style="@android:style/TextAppearance.Holo.Large"
        android:gravity="center" />

</LinearLayout>
```

在 res/layout 目录下新建一个名为 activity_second.xml 的布局文件，该文件中的代码如下：

```xml
<?xml version="1.0" encoding="utf-8"?>
<LinearLayout xmlns:android="http://schemas.android.com/apk/res/android"
    android:layout_width="match_parent"
    android:layout_height="match_parent"
    android:orientation="vertical" >

    <ImageView
        android:id="@+id/id_iv01"
        android:layout_width="match_parent"
        android:layout_height="0dp"
        android:layout_weight="1"
        android:scaleType="fitCenter"
        android:src="@drawable/png0001"
        android:contentDescription="@string/text_empty"/>

    <View
        android:layout_width="match_parent"
        android:layout_height="4dp"
        android:background="#aaa" />

    <ImageView
        android:id="@+id/id_iv02"
        android:layout_width="match_parent"
        android:layout_height="0dp"
        android:layout_weight="1"
        android:scaleType="fitCenter"
        android:src="@drawable/png0002"
```

```
            android:contentDescription="@string/text_empty"/>

</LinearLayout>
```

在 res/mipmap 目录下存储两个需要在布局文件 activity_second.xml 中用到的图片资源文件：png0001.png 文件和 png0002.png 文件。然后修改 res/values/strings.xml 文件中的代码，在其中定义布局文件中引用的字符串资源，修改后的代码如下：

```
<?xml version="1.0" encoding="utf-8"?>
<resources>

    <string name="app_name">Ch0702</string>

    <string name="text_button">点击</string>
    <string name="text_empty">""</string>

</resources>
```

修改 MainActivity.java 文件中的代码，使其显示 activity_main.xml 文件的界面，并且监听对按钮的点击事件，修改后的代码如下：

```
package com.example.ch0702;

import androidx.appcompat.app.AppCompatActivity;

import android.app.Activity;
import android.content.Intent;
import android.os.Bundle;
import android.view.View;
import android.widget.Button;
import android.widget.TextView;
import android.widget.Toast;

public class MainActivity extends AppCompatActivity
implements View.OnClickListener {
    public static int RC01 = 1000;

    private TextView tv;

    @Override
    protected void onCreate(Bundle savedInstanceState) {
        super.onCreate(savedInstanceState);
        setContentView(R.layout.activity_main);

        Button btn = this.findViewById(R.id.id_button);
```

```
        btn.setOnClickListener(this);

        tv = this.findViewById(R.id.id_textview);

    }

    @Override
    public void onClick(View v) {
        Intent intent = new Intent(this, SecondActivity.class);
        this.startActivityForResult(intent, MainActivity.RC01);
    }

    @Override
    protected void onActivityResult(int requestCode, int resultCode, Intent data) {
        super.onActivityResult(requestCode, resultCode, data);
        if ((requestCode == MainActivity.RC01) && (resultCode == Activity.RESULT_OK)) {
            int which = data.getIntExtra("result", -1);
            tv.setText(which + "");
        }
    }
}
```

在 onClick()方法中，使用 startActivityForResult()方法打开 SecondActivity，并且传递一个请求码 MainActivity.RC01，代码如下：

```
    @Override
    public void onClick(View v) {
        Intent intent = new Intent(this, SecondActivity.class);
        this.startActivityForResult(intent, MainActivity.RC01);
    }
```

当 SecondActivity 返回结果时，Android 会调用 MainActivity 的 onActivityResult()方法，在该方法中获取 SecondActivity 的返回结果，并且将其显示在 MainActivity 的 TextView 组件中。onActivityResult()方法中的代码如下：

```
    @Override
    protected void onActivityResult(int requestCode, int resultCode, Intent data) {
        super.onActivityResult(requestCode, resultCode, data);
        if ((requestCode == MainActivity.RC01) && (resultCode == Activity.RESULT_OK)) {
            int which = data.getIntExtra("result", -1);
```

```
            tv.setText(which + "");
        }
    }
```

在 src 目录下的 com.example.ch0702 包中新建一个名为 SecondActivity 的 Java 类文件，使其显示 activity_second.xml 文件的界面，并且添加对图片点击事件的响应方法。SecondActivity.java 文件中的代码如下：

```
package com.example.ch0702;

import android.app.Activity;
import android.content.Intent;
import android.os.Bundle;
import android.view.View;
import android.widget.ImageView;

public class SecondActivity extends Activity implements View.OnClickListener {

    @Override
    protected void onCreate(Bundle savedInstanceState) {
        super.onCreate(savedInstanceState);
        setContentView(R.layout.activity_second);

        ImageView iv01 = this.findViewById(R.id.id_iv01);
        iv01.setOnClickListener(this);
        ImageView iv02 = this.findViewById(R.id.id_iv02);
        iv02.setOnClickListener(this);
    }

    @Override
    public void onClick(View v) {
        int id = v.getId();
        Intent intent = new Intent();
        if (id == R.id.id_iv01) {
            intent.putExtra("result", 1);
        } else {
            intent.putExtra("result", 2);
        }
        this.setResult(Activity.RESULT_OK, intent);

        this.finish();
    }
}
```

在 SecondActivity 的 onClick()方法中，首先判断被点击的是哪张图片；然后创建一个 Intent

对象，用于给源 Activity 传递返回结果；最后调用 SecondActivity 的 finish()方法，用于关闭 SecondActivity。

在 AndroidManifest.xml 文件中配置 SecondActivity。修改 Android Manifest.xml 文件中的代码，修改后的代码如下：

```xml
<?xml version="1.0" encoding="utf-8"?>
<manifest xmlns:android="http://schemas.android.com/apk/res/android"
    package="com.example.ch0702">

    <application
        android:allowBackup="true"
        android:icon="@mipmap/ic_launcher"
        android:label="@string/app_name"
        android:roundIcon="@mipmap/ic_launcher_round"
        android:supportsRtl="true"
        android:theme="@style/Theme.Ch0702">
        <activity
            android:name=".MainActivity"
            android:exported="true">
            <intent-filter>
                <action android:name="android.intent.action.MAIN" />
                <category android:name="android.intent.category.LAUNCHER" />
            </intent-filter>
        </activity>

        <activity
            android:name=".SecondActivity"
            android:allowEmbedded="false">
        </activity>
    </application>

</manifest>
```

在修改完成后，运行 Ch0702 应用程序，即可得到图 7-4～图 7-6 所示的运行效果。

7.6 Intent 的综合应用案例

7.6.1 运行效果

本案例应用程序运行效果的首界面为一个功能列表，如图 7-7 所示。

点击"浏览网页"图片按钮，会启动浏览器，如图 7-8 所示。

点击"拨打电话"图片按钮，会显示拨打电话界面，如图 7-9 所示。

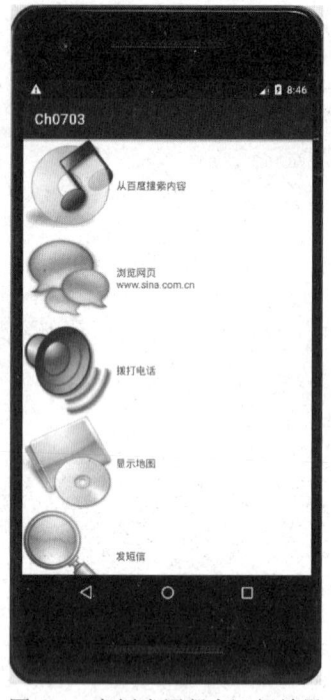

图 7-7 案例应用程序运行效果 图 7-8 启动浏览器 图 7-9 显示拨打电话界面
的首界面

7.6.2 程序代码

在 Android Studio 中新建一个名为 Ch0703 的 Android 应用程序工程，并且将需要用到的图片资源文件存储于 res/mipmap 目录下。修改布局文件 res/layout/activity_main.xml 中的代码，修改后的代码如下：

```xml
<?xml version="1.0" encoding="utf-8"?>
<ScrollView xmlns:android="http://schemas.android.com/apk/res/android"
    android:layout_width="match_parent"
    android:layout_height="match_parent">

    <TableLayout
        android:layout_width="match_parent"
        android:layout_height="wrap_content"
        android:stretchColumns="1" >

        <TableRow>
            <android.widget.Button
                android:id="@+id/button1"
                android:layout_width="wrap_content"
                android:layout_height="wrap_content"
                android:background="@mipmap/png0001" />
```

```xml
        <TextView
            android:layout_gravity="center_vertical"
            android:paddingLeft="4dp"
            android:paddingRight="4dp"
            android:text="@string/text_baidu" />
</TableRow>

<TableRow>

    <android.widget.Button
        android:id="@+id/button2"
        android:layout_width="wrap_content"
        android:layout_height="wrap_content"
        android:background="@mipmap/png0002" />

    <TextView
        android:layout_gravity="center_vertical"
        android:paddingLeft="4dp"
        android:paddingRight="4dp"
        android:text="@string/text_sina_web" />
</TableRow>

<TableRow>

    <android.widget.Button
        android:id="@+id/button3"
        android:layout_width="wrap_content"
        android:layout_height="wrap_content"
        android:background="@mipmap/png0003" />

    <TextView
        android:layout_gravity="center_vertical"
        android:paddingLeft="4dp"
        android:paddingRight="4dp"
        android:text="@string/text_dial" />
</TableRow>

<TableRow>

    <android.widget.Button
        android:id="@+id/button4"
        android:layout_width="wrap_content"
        android:layout_height="wrap_content"
        android:background="@mipmap/png0004" />

    <TextView
```

```xml
            android:layout_gravity="center_vertical"
            android:paddingLeft="4dp"
            android:paddingRight="4dp"
            android:text="@string/text_map" />
</TableRow>

<TableRow>
    <android.widget.Button
        android:id="@+id/button5"
        android:layout_width="wrap_content"
        android:layout_height="wrap_content"
        android:background="@mipmap/png0005" />

    <TextView
        android:layout_gravity="center_vertical"
        android:paddingLeft="4dp"
        android:paddingRight="4dp"
        android:text="@string/text_short_msg" />
</TableRow>

<TableRow>
    <android.widget.Button
        android:id="@+id/button6"
        android:layout_width="wrap_content"
        android:layout_height="wrap_content"
        android:background="@mipmap/png0006" />

    <TextView
        android:layout_gravity="center_vertical"
        android:paddingLeft="4dp"
        android:paddingRight="4dp"
        android:text="@string/text_email" />
</TableRow>

<TableRow>
    <android.widget.Button
        android:id="@+id/button7"
        android:layout_width="wrap_content"
        android:layout_height="wrap_content"
        android:background="@mipmap/png0007" />

    <TextView
        android:layout_gravity="center_vertical"
```

```xml
                android:paddingLeft="4dp"
                android:paddingRight="4dp"
                android:text="@string/text_multimedia" />
        </TableRow>
    </TableLayout>

</ScrollView>
```

在上述代码中，使用 TableLayout 容器组件显示每行的按钮和文字，由于表格中所有行的总高度可能会大于物理屏幕的高度，因此在 TableLayout 容器组件的外面设置了一个 ScrollView 组件，用于实现垂直滚动功能。在一般情况下，在比较规整的界面布局中，可以使用 TableLayout 容器组件与 TableRow 组件相结合的方式实现行列的规整布局：在 TableLayout 容器组件中包含 TableRow 组件，其中，TableRow 组件表示表格中的一行。TableLayout 容器组件的几个典型 XML 配置参数及其含义如下。

- android:stretchColumns。当显示一行数据时，如果物理屏幕在宽度上有富余的空间，则将指定列的宽度自动延展，以便使用富余的显示空间。
- android:shrinkColumns。当显示一行数据时，如果物理屏幕的宽度不足以显示所有列的数据，则自动压缩指定列的宽度，使其他列有足够的显示空间。
- android:collapseColumns。不显示指定的列。

以上 3 个属性都支持以英文逗号"，"分隔多个列号，列号从 0 开始。

修改 res/values/strings.xml 文件中的代码，在其中定义布局文件中引用的字符串资源，修改后的代码如下：

```xml
<?xml version="1.0" encoding="utf-8"?>
<resources>

    <string name="app_name">Ch0703</string>

    <string name="text_baidu">从百度搜索内容</string>
    <string name="text_sina_web">浏览网页\nwww.sina.com.cn</string>
    <string name="text_dial">拨打电话</string>
    <string name="text_map">显示地图</string>
    <string name="text_short_msg">发短信</string>
    <string name="text_email">发送 Email</string>
    <string name="text_multimedia">播放多媒体</string>

</resources>
```

修改 MainActivity.java 文件中的代码，使其显示 activity_main.xml 文件的界面，监听对按钮的点击事件，并且根据所点击的按钮创建相应的 Intent 对象，用于启动并运行相应的 Activity。需要注意的是，使用 Intent 对象启动的 Activity 可以是当前应用程序的 Activity，也可以是第三方应用程序的 Activity。MainActivity.java 文件中的代码如下：

```
package com.example.ch0703;
```

```java
import androidx.appcompat.app.AppCompatActivity;

import android.app.SearchManager;
import android.content.Intent;
import android.net.Uri;
import android.os.Bundle;
import android.provider.MediaStore;
import android.view.View;
import android.widget.Button;

public class MainActivity extends AppCompatActivity implements View.OnClickListener {
    Button btn1,btn2,btn3,btn4,btn5,btn6,btn7;

    @Override
    protected void onCreate(Bundle savedInstanceState) {
        super.onCreate(savedInstanceState);
        setContentView(R.layout.activity_main);

        btn1 = this.findViewById(R.id.button1);
        btn1.setOnClickListener(this);

        btn2 = this.findViewById(R.id.button2);
        btn2.setOnClickListener(this);

        btn3 = this.findViewById(R.id.button3);
        btn3.setOnClickListener(this);

        btn4 = this.findViewById(R.id.button4);
        btn4.setOnClickListener(this);

        btn5 = this.findViewById(R.id.button5);
        btn5.setOnClickListener(this);

        btn6 = this.findViewById(R.id.button6);
        btn6.setOnClickListener(this);

        btn7 = this.findViewById(R.id.button7);
        btn7.setOnClickListener(this);
    }
```

```java
@Override
public void onClick(View v) {
    Uri uri;
    Intent it;
    switch(v.getId()) {
        case R.id.button1:
            Intent intent = new Intent();
            intent.setAction(Intent.ACTION_WEB_SEARCH);
            intent.putExtra(SearchManager.QUERY,"Andoid应用编程");
            startActivity(intent);
            break;
        case R.id.button2:
            uri = Uri.parse("http://www.sina.com.cn");
            it = new Intent(Intent.ACTION_VIEW,uri);
            startActivity(it);
            break;
        case R.id.button3:
            uri = Uri.parse("tel:5556");
            it = new Intent(Intent.ACTION_DIAL, uri);
            startActivity(it);
            break;
        case R.id.button4:
            uri = Uri.parse("geo:38.899533,-77.036476");
            it = new Intent(Intent.ACTION_VIEW,uri);
            startActivity(it);
            break;
        case R.id.button5:
            uri = Uri.parse("smsto:5556");
            it = new Intent(Intent.ACTION_SENDTO, uri);
            it.putExtra("sms_body", "The SMS text");
            startActivity(it);
            break;
        case R.id.button6:
            uri = Uri.parse("mailto:xxx@abc.com");
            it = new Intent(Intent.ACTION_SENDTO, uri);
            startActivity(it);
            break;
        case R.id.button7:
            uri = Uri.withAppendedPath(
                    MediaStore.Audio.Media.INTERNAL_CONTENT_URI, "1");
```

```
                it = new Intent(Intent.ACTION_VIEW, uri);
                startActivity(it);
                break;
        }
    }
}
```

7.7 同步练习二

完善 7.6 节中的案例应用程序，要求如下。
- 将显示在按钮右边的字体调大，使其与左边的按钮大小相匹配。
- 新建一个 Activity，点击图 7-7 所示界面中的一个按钮，能够启动并运行该 Activity，该 Activity 会返回一些数据给 MainActivity，并且 MainActivity 能够显示返回的数据。

7.8 广播消息和广播接收器

使用 Intent 对象不仅可以启动新的 Activity，还可以在应用程序之间广播消息。下面介绍如何使用 Intent 对象广播消息，以及如何接收 Android 平台的广播消息，以便应用程序对系统状态的变化做出响应。

Android 提供了两种可广播消息，即普通消息和有序消息。普通消息是指所有对该消息感兴趣的接收者都可以收到的消息，接收者收到该消息的顺序是随机的。有序消息是指按指定优先级传递的消息，将消息先传递给具有最高优先级的接收者，具有最高优先级的接收者可以决定是否继续转发该消息给其他低优先级的接收者。我们只介绍普通消息，对于发送和接收有序消息，以及为消息设置接收权限的相关知识，读者可以参考 Android SDK 帮助文档。

7.8.1 发送和接收普通消息

所有 Activity 都可以使用 sendBroadcast(Intent intent)方法和 sendBroadcast(Intent intent, String receiverPermission)方法发送普通消息。其中，使用 sendBroadcast(Intent intent)方法可以将普通消息发送给所有的接收者，无论接收者是否具有接收权限；使用 sendBroadcast(Intent intent, String receiverPermission)方法可以将普通消息发送给具有接收权限的接收者。

要接收 sendBroadcast()方法发送的普通消息，接收者必须继承 BroadcastReceiver 类，并且需要实现其中的 onReceive(Context context, Intent intent)方法。此外，需要在 AndroidManifest.xml 文件或程序代码中注册这个接收者。

下面举例说明如何发送和接收普通消息。该案例应用程序的首界面中有一个输入框和一个按钮，在输入框中输入要发送的消息，点击按钮即可将消息发送出去；消息接收者在接收到该消息后，会通过 Toast 组件将该消息显示出来。该案例应用程序运行效果的首界面如图 7-10 所示，在输入框中输入一个字符串，然后点击"发送"按钮，将消息发送出去，即可在 Toast 组件中显示接收到的消息，如图 7-11 所示。

图 7-10　案例应用程序运行效果的首界面　　图 7-11　在 Toast 组件中显示接收到的消息

在 Android Studio 中新建一个名为 Ch0704 的 Android 应用程序工程。

修改布局文件 res/layout/activity_main.xml 中的代码，修改后的代码如下：

```xml
<?xml version="1.0" encoding="utf-8"?>
<LinearLayout xmlns:android="http://schemas.android.com/apk/res/android"
    android:layout_width="match_parent"
    android:layout_height="match_parent"
    android:orientation="vertical">

    <TextView
        android:layout_width="match_parent"
        android:layout_height="wrap_content"
        android:text="@string/text_input_something" />

    <EditText
        android:id="@+id/id_edittext"
        android:layout_width="match_parent"
        android:layout_height="200dp"
        android:singleLine="false"
        android:inputType="text"/>

    <Button
        android:id="@+id/id_button"
```

```xml
            android:layout_width="match_parent"
            android:layout_height="wrap_content"
            android:text="@string/text_button"/>

</LinearLayout>
```

这个布局文件很简单,只在 LinearLayout 容器组件中放置了几个基本组件。

修改 res/values/strings.xml 文件中的代码,在其中定义布局文件中引用的字符串资源,修改后的代码如下:

```xml
<resources>
    <string name="app_name">Ch0704</string>

    <string name="text_input_something">请输入要发送的消息:</string>
    <string name="text_button">发送</string>
</resources>
```

修改 MainActivity.java 文件中的代码,使其显示 activity_main.xml 文件的界面,监听对按钮的点击事件,并且在 onClick()方法中发送普通消息,修改后的代码如下:

```java
package com.example.ch0704;

import androidx.appcompat.app.AppCompatActivity;

import android.content.Intent;
import android.os.Bundle;
import android.view.View;
import android.widget.Button;
import android.widget.EditText;

public class MainActivity extends AppCompatActivity implements View.OnClickListener {
    private EditText et;

    @Override
    protected void onCreate(Bundle savedInstanceState) {
        super.onCreate(savedInstanceState);
        setContentView(R.layout.activity_main);

        et = (EditText)this.findViewById(R.id.id_edittext);
        Button btn = (Button)this.findViewById(R.id.id_button);
        btn.setOnClickListener(this);

    }

    @Override
```

```java
    public void onClick(View v) {
        String msg = et.getText().toString();
        
        Intent intent = new Intent("com.example.ch0704.hello");
        intent.putExtra("msg", msg);
        this.sendBroadcast(intent);
    }

}
```

在 onClick()方法的实现代码中创建一个 Intent 对象,并且在该对象中存储相应的消息,然后广播该消息。

下面需要编写一个消息接收器,用于接收该消息。在 src 目录下的 com.example.ch0704 包中新建一个名为 MyReceiver 的 Java 类文件,并修改该文件中的代码,修改后的代码如下:

```java
package com.example.ch0704;

import android.content.BroadcastReceiver;
import android.content.Context;
import android.content.Intent;
import android.widget.Toast;

public class MyReceiver extends BroadcastReceiver {

    @Override
    public void onReceive(Context context, Intent intent) {
        String msg = intent.getStringExtra("msg");
        Toast.makeText(context, msg, Toast.LENGTH_LONG).show();
    }

}
```

作为消息接收器,MyReceiver 类必须继承 BroadcastReceiver 类,并且重写其中的 onReceive()方法,使用 Toast 组件将接收到的消息显示出来。

在 AndroidManifest.xml 文件中注册该消息接收器。修改 AndroidManifest.xml 文件中的代码,修改后的代码如下:

```xml
<?xml version="1.0" encoding="utf-8"?>
<manifest xmlns:android="http://schemas.android.com/apk/res/android"
    package="com.example.ch0704">

    <application
        android:allowBackup="true"
        android:icon="@mipmap/ic_launcher"
        android:label="@string/app_name"
        android:roundIcon="@mipmap/ic_launcher_round"
```

```xml
            android:supportsRtl="true"
            android:theme="@style/Theme.Ch0704">
            <activity
                android:name=".MainActivity"
                android:exported="true">
                <intent-filter>
                    <action android:name="android.intent.action.MAIN" />

                    <category android:name="android.intent.category.LAUNCHER" />
                </intent-filter>
            </activity>

            <receiver android:name=".MyReceiver"
                android:exported="false">
                <intent-filter>
                    <action android:name="com.example.ch0704.hello"/>
                </intent-filter>
            </receiver>

        </application>

</manifest>
```

注意以下代码片段。

```xml
            <receiver android:name=".MyReceiver"
                android:exported="false">
                <intent-filter>
                    <action android:name="com.example.ch0704.hello"/>
                </intent-filter>
            </receiver>
```

上述代码片段注册了 MyReceiver 消息接收器，使其可以监听 action 中 name 为"com.example.ch0704.hello"的消息。

运行 Ch0704 应用程序，即可得到图 7-10 所示的界面，在输入框中输入一些文字，点击"发送"按钮，即可将消息发送给 MyReceiver 消息接收器，并且在 MyReceiver 消息接收器中使用 Toast 组件将接收到的消息显示出来。

除了可以在 AndroidManifest.xml 文件中注册消息接收者（称为静态注册），还可以在 Java 代码中注册消息接收者（称为动态注册）。下面，从 AndroidManifest.xml 文件中删除对 MyReceiver 消息接收器的注册，然后在 MainActivity.java 文件中动态注册 MyReceiver 消息接收器。修改后的 MainActivity.java 文件中的代码如下：

```java
package com.example.ch0704;

import androidx.appcompat.app.AppCompatActivity;
```

```java
import android.content.Intent;
import android.content.IntentFilter;
import android.os.Bundle;
import android.view.View;
import android.widget.Button;
import android.widget.EditText;

public class MainActivity extends AppCompatActivity implements View.OnClickListener {
    private EditText et;
    private MyReceiver receiver;

    @Override
    protected void onCreate(Bundle savedInstanceState) {
        super.onCreate(savedInstanceState);
        setContentView(R.layout.activity_main);

        receiver = new MyReceiver();

        et = (EditText)this.findViewById(R.id.id_edittext);
        Button btn = (Button)this.findViewById(R.id.id_button);
        btn.setOnClickListener(this);

    }

    @Override
    public void onClick(View v) {
        String msg = et.getText().toString();

        Intent intent = new Intent("com.example.ch0704.hello");
        intent.putExtra("msg", msg);
        this.sendBroadcast(intent);
    }

    @Override
    protected void onResume() {
        super.onResume();

        IntentFilter filter = new IntentFilter();
        filter.addAction("com.example.ch0704.hello");
        registerReceiver(receiver, filter);
    }
```

```
@Override
protected void onPause() {
    super.onPause();

    unregisterReceiver(receiver);
}
}
```

在修改后的 MainActivity.java 文件中，在 onResume()生命周期方法中注册 MyReceiver 消息接收器，并且在 onPause()生命周期方法中解除对广播的监听。运行修改后的 Ch0704 应用程序，也可以得到图 7-10 所示的界面。

7.8.2 接收 Android 平台广播的普通消息

Android 平台在特定的情况下会广播特定的普通消息，任何感兴趣的接收者通过注册均可以接收到这些普通消息。Android 平台常用的普通消息及其含义如表 7-1 所示。

表 7-1 Android 平台常用的普通消息及其含义

普通消息名称	含 义
Intent.ACTION_TIME_TICK	每一分钟发送一次消息
Intent.ACTION_BOOT_COMPLETED	在 Android 完成启动时发送该消息
Intent.ACTION_SHUTDOWN	在关机时发送该消息
Intent.ACTION_BATTERY_LOW	在电池的电量低于指定值时发送该消息
Intent.ACTION_POWER_CONNECTED	在连接电源时发送该消息
Intent.ACTION_POWER_DISCONNECTED	在断开电源连接时发送该消息

7.9 同步练习三

编写一个消息接收器应用程序，该应用程序既可以接收自己编写的应用程序发送的普通消息，又可以接收 Android 平台发送的 Intent.ACTION_POWER_CONNECTED 消息，设计界面并测试该应用程序。

第 8 章 构建菜单应用程序

菜单是一种常见的应用程序操作模式,早期的 Android 设备提供了专门用于开启菜单的功能按钮,但是,从 Android 3.0 开始,Android 不再要求设备制造商提供这个功能按钮,改为使用应用程序顶部的菜单弹出按钮(3 个小点)打开菜单。本章主要介绍如何构建 Android 菜单应用程序。

8.1 菜单

在使用 Android Studio 构建的 Android 应用程序界面中,会显示一个菜单弹出按钮(方框框住部分),如图 8-1 所示。点击这个按钮,会显示菜单功能项,如图 8-2 所示。

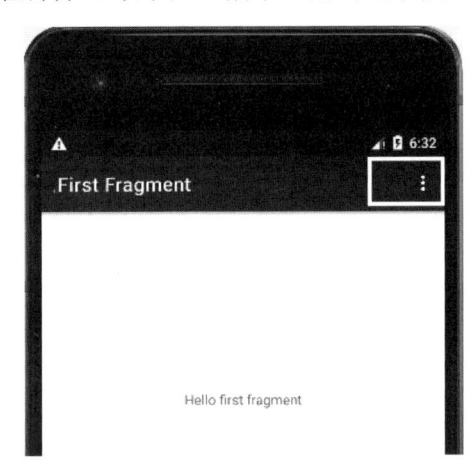

图 8-1 菜单弹出按钮 　　　　　　　　图 8-2 菜单功能项

菜单功能项中可以显示更多的选项。下面举例说明 Android 菜单的使用方法。

在 Android Studio 中新建一个名为 Ch0801 的 Android 应用程序工程,在 res 目录下创建 menu 子目录,然后创建 res/menu/menu_main.xml 文件,修改该文件中的代码,修改后的代码如下:

```xml
<menu xmlns:android="http://schemas.android.com/apk/res/android"
    xmlns:app="http://schemas.android.com/apk/res-auto"
    xmlns:tools="http://schemas.android.com/tools"
    tools:context="com.example.ch0801.MainActivity">

    <item
```

```xml
        android:id="@+id/id_mi_phone"
        android:icon="@mipmap/png0001"
        android:orderInCategory="100"
        app:showAsAction="ifRoom|withText"
        android:title="@string/text_mi_phone"/>

    <item
        android:id="@+id/id_mi_cut"
        android:icon="@mipmap/png0002"
        android:orderInCategory="100"
        app:showAsAction="ifRoom|withText"
        android:title="@string/text_mi_cut"/>

    <item
        android:id="@+id/id_mi_trump"
        android:icon="@mipmap/png0003"
        android:orderInCategory="100"
        app:showAsAction="ifRoom|withText"
        android:title="@string/text_mi_trump"/>

    <item
        android:id="@+id/id_mi_book"
        android:icon="@mipmap/png0004"
        android:orderInCategory="100"
        app:showAsAction="ifRoom|withText"
        android:title="@string/text_mi_book"/>

    <item
        android:id="@+id/id_mi_save"
        android:icon="@mipmap/png0005"
        android:orderInCategory="99"
        app:showAsAction="ifRoom|withText"
        android:title="@string/text_mi_save"/>

    <item
        android:id="@+id/id_mi_mail"
        android:icon="@mipmap/png0006"
        android:orderInCategory="100"
        app:showAsAction="ifRoom|withText"
        android:title="@string/text_mi_mail"/>

</menu>
```

在 res/mipmap 目录下存储图片资源文件 png0001.png～png0006.png。

这个菜单中共包括 6 个菜单功能项，并且为每个菜单功能项都指定了 id、图标和标题。在每个菜单功能项中，android:orderInCategory 属性主要用于指定将当前菜单功能项显示在菜单栏中的优先级，值越小，优先级越高；app:showAsAction 属性主要用于指定是否将当前菜单功能项显示在菜单栏中，其值包括 never、ifRoom、withText 和 always，可以对这些值进行或运算。此外，需要修改 res/values/strings.xml 文件中的代码，在其中定义布局文件中引用的字符串资源，修改后的代码如下：

```xml
<resources>
    <string name="app_name">Ch0801</string>

    <string name="text_mi_phone">电话</string>
    <string name="text_mi_cut">剪切</string>
    <string name="text_mi_trump">喇叭</string>
    <string name="text_mi_book">书籍</string>
    <string name="text_mi_save">保存</string>
    <string name="text_mi_mail">邮件</string>
</resources>
```

为了能够在菜单栏中显示菜单功能项，或者在点击 MENU 按钮或菜单弹出按钮时显示菜单功能项，需要重写 MainActivity 类中的 onCreateOptionsMenu()方法；为了响应对菜单功能项的选择事件，需要重写 MainActivity 类中的 onOptionsItemSelected()方法。修改后的 MainActivity.java 文件中的代码如下：

```java
package com.example.ch0801;

import androidx.appcompat.app.AppCompatActivity;

import android.os.Bundle;
import android.view.Menu;
import android.view.MenuItem;
import android.widget.Toast;

public class MainActivity extends AppCompatActivity {

    @Override
    protected void onCreate(Bundle savedInstanceState) {
        super.onCreate(savedInstanceState);
        setContentView(R.layout.activity_main);
    }

    @Override
    public boolean onCreateOptionsMenu(Menu menu) {
```

```java
            getMenuInflater().inflate(R.menu.menu_main, menu);
        return true;
    }


    @Override
    public boolean onOptionsItemSelected(MenuItem item) {
        int id = item.getItemId();
        switch(id) {
            case R.id.id_mi_phone:
                Toast.makeText(this, "启动打电话", Toast.LENGTH_LONG).show();
                break;
            case R.id.id_mi_cut:
                Toast.makeText(this, "剪切", Toast.LENGTH_LONG).show();
                break;
            case R.id.id_mi_trump:
                Toast.makeText(this, "吹喇叭", Toast.LENGTH_LONG).show();
                break;
            case R.id.id_mi_book:
                Toast.makeText(this, "书籍", Toast.LENGTH_LONG).show();
                break;
            case R.id.id_mi_save:
                Toast.makeText(this, "假装的保存", Toast.LENGTH_LONG).show();
                break;
            case R.id.id_mi_mail:
                Toast.makeText(this, "启动发邮件功能", Toast.LENGTH_LONG).show();
                break;
        }
        return super.onOptionsItemSelected(item);
    }
}
```

注意 onCreateOptionsMenu()方法。当用户点击手机上的 MENU 按钮或点击应用程序顶部的菜单弹出按钮时，会自动调用该方法，用于生成所需的菜单，因此，在该方法中，使 getMenuInflater()方法获取一个菜单展开器，使用这个菜单展开器打开 res/menu 目录下定义的菜单文件 menu_main.xml。

注意 onOptionsItemSelected()方法。该方法是在用户选择菜单功能项时被调用的，可以根据 item 对象中被选中菜单功能项的 id 参数识别被选中的是哪个菜单功能项，从而执行相应的操作。

运行 Ch0801 应用程序，运行效果如图 8-3 所示。

点击菜单弹出按钮，会显示菜单功能项，如图 8-4 所示，选择任意一个菜单功能项，都会使用 Toast 组件显示一个提示信息，用于表示该菜单功能项是工作的，读者也可以修改菜单功能项的功能代码，使其完成其他工作。

图 8-3　Ch0801 应用程序的运行效果

图 8-4　显示菜单功能项

8.2　同步练习

编写一个简单的菜单应用程序，在菜单中显示两个菜单功能项，用于切换不同的图片：选择"上一张"菜单功能项，会显示上一张图片；选择"下一张"菜单功能项，会显示下一张图片。

第 9 章

动画

Android 支持两种类型的动画,分别为属性动画和 View 动画。View 动画分为两种,分别为补间动画和帧动画。动画的基本原理是在特定的时间内,将组件的某个属性值从一个值变化到一个新的值,或者将某个组件的显示状态从一种状态变化到一种新的状态。

View 动画是最常用的动画,因此,本章主要介绍 View 动画的补间动画和帧动画。

9.1 View 动画之补间动画基础

9.1.1 补间动画举例

补间动画的基本原理是在特定的时间内,按照一定的速度,将组件从一个位置移动到另一个位置、从一个角度旋转到另一个角度、显示透明度从一个值变换到另一个值、大小从一个尺寸变换到另一个尺寸,Android 还允许将以上 4 种操作合并在一个动画中执行。

下面来看一个简单的案例:在 4 秒内,以一个 TextView 组件的左下角为旋转中心,将其从 0 度顺时针旋转到 90 度,并且旋转的速度越来越快。该案例应用程序的运行效果如图 9-1 所示,TextView 组件会持续旋转,直到 4 秒后完成动画并显示最终效果。

下面构建该案例应用程序。首先在 Android Studio 中新建一个名为 Ch0901 的 Android 应用程序工程,然后修改布局文件 res/layout/activity_main.xml 中的代码,修改后的代码如下:

```xml
<RelativeLayout xmlns:android="http://schemas.android.com/apk/res/android"
    xmlns:tools="http://schemas.android.com/tools"
    android:layout_width="match_parent"
    android:layout_height="match_parent">

    <TextView
        android:id="@+id/id_tv"
        android:layout_width="wrap_content"
        android:layout_height="wrap_content"
        android:text="@string/hello_world" />

</RelativeLayout>
```

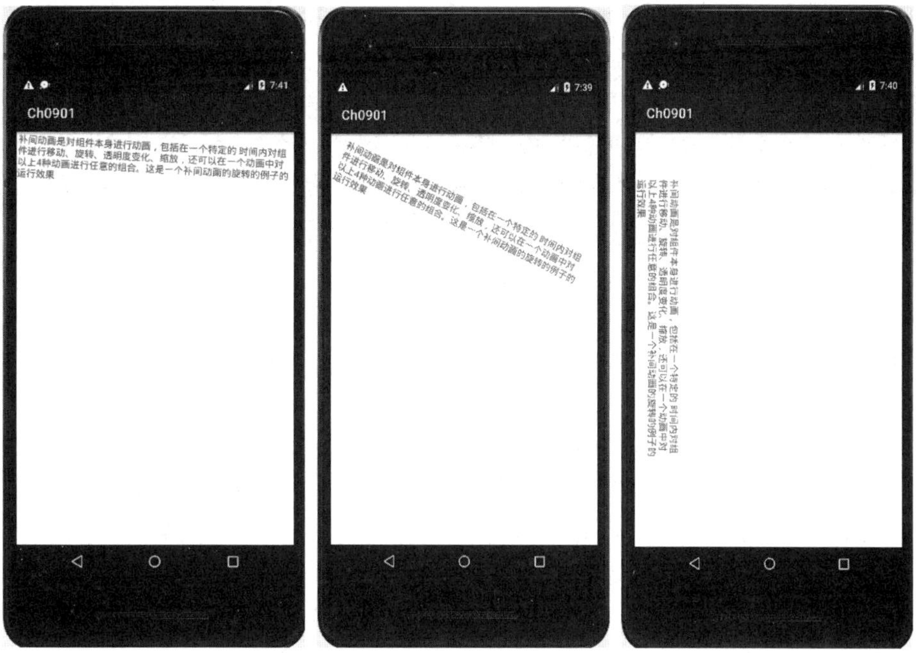

图 9-1　旋转 TextView 组件案例应用程序的运行效果

为了使 TextView 组件显示所需文字，修改 res/values/strings.xml 文件中的代码，修改后的代码如下：

```
<resources>
    <string name="app_name">Ch0901</string>

    <string name="hello_world">补间动画是对组件本身进行动画，包括在一个特定的
        时间内对组件进行移动、旋转、透明度变化、缩放，还可以在一个动画中对
        以上 4 种动画进行任意的组合。这是一个补间动画的旋转的例子的运行效果
    </string>

</resources>
```

Android 提供了两种定义动画的方式：使用 XML 文件定义动画和使用 Java 代码定义动画。Android 建议使用 XML 文件定义动画，并且 View 动画定义文件必须存储于工程的 res/anim 目录下。因此，在工程的 res 目录下创建 anim 子目录，并且在 res/anim 目录下创建一个名为 my_rotate.xml 的 View 动画定义文件。修改 my_rotate.xml 文件中的代码，修改后的代码如下：

```
<?xml version="1.0" encoding="utf-8"?>
<rotate xmlns:android="http://schemas.android.com/apk/res/android"
    android:interpolator="@android:anim/accelerate_interpolator"
    android:fromDegrees="0"
    android:toDegrees="90"
    android:pivotX="0%"
    android:pivotY="100%"
    android:duration="4000"
```

```
android:fillAfter="true">
</rotate>
```

在上述代码中，<rotate>标签表示要定义旋转动画。以下代码表示旋转动画的执行方式，称为动画插值器，其值为 accelerate_interpolator，表示动画的执行速度会越来越快。

```
android:interpolator="@android:anim/accelerate_interpolator"
```

以下代码表示旋转的初始角度和结束角度，分别为 0 度和 90 度。

```
android:fromDegrees="0"
android:toDegrees="90"
```

以下代码表示旋转中心的坐标，其中 android:pivotX 表示旋转中心的横坐标，android:pivotY 表示旋转中心的纵坐标。

```
android:pivotX="0%"
android:pivotY="100%"
```

有两种表示旋转中心坐标的方式：带%的值和不带%的值。带%的值表示以组件的长度或高度为度量基础，不带%的值表示以像素为度量基础。在不知道组件实际大小的情况下，使用带%的值会更容易一些。例如，在上述代码中，我们指定旋转中心的横坐标偏移组件长度的 0%，也就是 $x=0px$；指定旋转中心的纵坐标偏移组件高度的 100%，也就是 $y=$ 组件的高度。因此，上述代码表示旋转中心为组件左下角的点。

以下代码定义了动画执行的时间，单位是毫秒。

```
android:duration="4000"
```

以下代码表示动画执行完毕后组件的显示状态，值为 true 表示组件保持动画执行完毕后的状态，值为 false 表示组件在动画执行完毕后恢复动画执行前的状态。

```
android:fillAfter="true">
```

在定义好动画文件后，接下来需要修改 MainActivity.java 文件中的代码，使 TextView 组件可以被动画处理，修改后的代码如下：

```
package com.example.ch0901;

import androidx.appcompat.app.AppCompatActivity;

import android.os.Bundle;
import android.view.animation.Animation;
import android.view.animation.AnimationUtils;
import android.widget.TextView;

public class MainActivity extends AppCompatActivity {

    @Override
    protected void onCreate(Bundle savedInstanceState) {
        super.onCreate(savedInstanceState);
        setContentView(R.layout.activity_main);
```

```
        TextView tv = (TextView) findViewById(R.id.id_tv);
        Animation rotate = AnimationUtils.loadAnimation(this,
R.anim.my_rotate);
        tv.startAnimation(rotate);
    }
}
```

在 onCreate()回调方法中，首先获取 TextView 组件，然后使用 AnimationUtils 类加载动画资源，最后在 TextView 组件上执行动画，代码如下：

```
        TextView tv = (TextView) findViewById(R.id.id_tv);
        Animation rotate = AnimationUtils.loadAnimation(this, R.anim.my_rotate);
        tv.startAnimation(rotate);
```

运行 Ch0901 应用程序，即可得到图 9-1 所示的运行效果。

9.1.2　补间动画的形式

补间动画包括 5 种动画形式，分别为旋转动画、缩放动画、透明度动画、移位动画和复合动画，标签分别为<rotate>、<scale>、<alpha>、<translate>和<set>。定义补间动画的文件必须存储于工程的 res/anim 目录下，文件名称可以是任意合法的名称，在 Java 程序代码中采用 R.anim.filename 的方式引用动画资源，在 XML 文件中采用@[package:]anim/filename 的方式引用动画资源。

1．定义旋转动画

定义旋转动画的一般格式如下：

```
<?xml version="1.0" encoding="utf-8"?>
<rotate xmlns:android="http://schemas.android.com/apk/res/android"
    android:interpolator="@[package:]anim/interpolator_resource"
    android:fromDegrees="float"
    android:toDegrees="float"
    android:pivotX="float"
    android:pivotY="float"
    android:startOffset="int"
    android:duration="int"
    android:fillAfter="boolean"
    android:repeatCount="int"
    android:repeatMode="restart | reverse"
    >
</rotate>
```

以下代码表示动画插值器，也就是采用何种方式改变组件的属性值或组件的显示状态。

```
android:interpolator="@[package:]anim/interpolator_resource"
```

Android 内置的动画插值器如图 9-2 所示。

每个动画插值器的表现形式很难用语言表达清楚，读者可以使用图 9-2 中的动画插值器替换 9.1.1 节案例应用程序中的 accelerate_interpolator 动画插值器，从而查看每个动画插值器的表现形式。

```
@android:anim/accelerate_decelerate_interpolator
@android:anim/accelerate_interpolator
@android:anim/anticipate_interpolator
@android:anim/anticipate_overshoot_interpolator
@android:anim/bounce_interpolator
@android:anim/cycle_interpolator
@android:anim/decelerate_interpolator
@android:anim/fade_in
@android:anim/fade_out
@android:anim/linear_interpolator
@android:anim/overshoot_interpolator
@android:anim/slide_in_left
@android:anim/slide_out_right
```

图 9-2　Android 内置的动画插值器

以下代码分别表示旋转的初始角度、最终角度和旋转中心坐标，这些参数已经在 9.1.1 节中介绍过。

```
android:fromDegrees="float"
android:toDegrees="float"
android:pivotX="float"
android:pivotY="float"
```

以下代码分别表示启动动画的时间延迟和执行动画的时间。

```
android:startOffset="int"
android:duration="int"
```

以下代码表示动画执行完毕后组件的显示状态。

```
android:fillAfter="boolean"
```

以下代码分别表示动画重复的次数和动画重复的方式。动画重复的方式可取值为 restart 和 reverse，restart 表示再次执行动画，reverse 表示以相反的方式再次执行动画。

```
android:repeatCount="int"
android:repeatMode="restart | reverse"
```

2. 定义缩放动画

定义缩放动画的一般格式如下：

```
<?xml version="1.0" encoding="utf-8"?>
<scale xmlns:android="http://schemas.android.com/apk/res/android"
    android:interpolator="@[package:]anim/interpolator_resource"
    android:fromXScale="float"
    android:toXScale="float"
    android:fromYScale="float"
    android:toYScale="float"
    android:pivotX="float"
    android:pivotY="float"
    android:startOffset="int"
```

```
        android:duration="int"
        android:fillAfter="boolean"
        android:repeatCount="int"
        android:repeatMode="restart | reverse"
        >
</scale>
```

以下代码表示组件在 x 轴方向上以组件宽度为基础的组件宽度变换比例。

```
        android:fromXScale="float"
        android:toXScale="float"
```

例如,以下代码表示宽度从 1.4 倍变换到 0.5 倍。

```
android:fromXScale="1.4"
android:toXScale="0.5"
```

以下代码表示组件在 y 轴方向上以组件高度为基础的组件高度变换比例。

```
        android:fromYScale="float"
        android:toYScale="float"
```

其他参数与定义旋转动画的相关参数含义相同。

3. 定义透明度动画

定义透明度动画的一般格式如下:

```
<?xml version="1.0" encoding="utf-8"?>
<alpha xmlns:android="http://schemas.android.com/apk/res/android"
    android:interpolator="@[package:]anim/interpolator_resource"
    android:fromAlpha="float"
    android:toAlpha="float"
    android:startOffset="int"
    android:duration="int"
    android:fillAfter="boolean"
    android:repeatCount="int"
    android:repeatMode="restart | reverse"
    >
</alpha>
```

以下代码表示组件的透明度值的变化,取值范围为 0.0～1.0,其中,0.0 表示完全透明,1.0 表示完全不透明。

```
        android:fromAlpha="float"
        android:toAlpha="float"
```

其他参数与定义旋转动画的相关参数含义相同。

4. 定义移位动画

定义移位动画的一般格式如下:

```
<?xml version="1.0" encoding="utf-8"?>
<translate xmlns:android="http://schemas.android.com/apk/res/android"
    android:interpolator="@[package:]anim/interpolator_resource"
    android:fromXDelta="float"
```

```
        android:fromYDelta="float"
        android:toXDelta="float"
        android:toYDelta="float"
        android:startOffset="int"
        android:duration="int"
        android:fillAfter="boolean"
        android:repeatCount="int"
        android:repeatMode="restart | reverse"
        >
</translate>
```

以下代码表示组件左上角的移位起始点坐标，android:fromXDelta 表示移位起始点的横坐标，android:fromYDelta 表示移位起始点的纵坐标。

```
        android:fromXDelta="float"
        android:fromYDelta="float"
```

移位起始点坐标的表示方式有 3 种：普通数值、带%的数值和带%p 的数值。其中，普通数值表示像素点坐标值，如 android:fromXDelta="5"表示移位起始点的横坐标为 5px；带%的数值表示以组件的宽度或高度为度量基础的像素点坐标值，如 android:fromXDelta="5%"表示移位起始点的横坐标为组件宽度的 5%；带%p 的数值表示以组件的父容器组件的宽度或高度为度量基础的像素点坐标值。

以下代码表示组件左上角的移位终止点坐标。移位终止点坐标的表示方式与移位起始点坐标的表示方式相同。

```
        android:toXDelta="float"
        android:toYDelta="float"
```

5．定义复合动画

除了可以使用<rotate>、<scale>、<alpha>和<translate>标签定义动画，还可以使用<set>标签定义复合动画。复合动画是指将多个动画合并成一个动画。定义复合动画的一般格式如下：

```
<?xml version="1.0" encoding="utf-8"?>
<set xmlns:android="http://schemas.android.com/apk/res/android"
     android:interpolator="@[package:]anim/interpolator_resource"
     android:shareInterpolator=["true" | "false"] >
  <rotate …/>
  <translate …/>
  …
  <set …> … </set>
</set>
```

复合动画是其他 4 种动画的组合，并且复合动画中可以包含复合动画。

注意以下配置参数。

```
        android:shareInterpolator=["true" | "false"]>
```

该配置参数主要表示在复合动画中指定的动画插值器是否作用于所有在该复合动画中定义的子动画，值为 true 表示是，值为 false 表示否。

9.1.3 使用动画监听器接口

使用动画监听器接口 Animation.AnimationListener 可以监听动画执行的各个阶段。查看 Android 帮助文档可知,动画监听器接口 Animation.AnimationListener 中包含 3 个方法,如图 9-3 所示。这 3 个方法分别用于监听动画的开始（onAnimationStart()）、结束（onAnimationEnd()）和重复（onAnimationRepeat()）。

Public Methods	
abstract void	onAnimationEnd(Animation animation) Notifies the end of the animation.
abstract void	onAnimationRepeat(Animation animation) Notifies the repetition of the animation.
abstract void	onAnimationStart(Animation animation) Notifies the start of the animation.

图 9-3　动画监听器接口 Animation.AnimationListener 中的 3 个方法

修改 9.1.1 节中的 Ch0901 应用程序,在旋转动画执行完毕后,使用 Toast 组件显示一条结束消息。为此,我们只需修改 MainActivity.java 文件中的 onCreate()回调方法,修改后的 onCreate()回调方法如下:

```java
@Override
protected void onCreate(Bundle savedInstanceState) {
    super.onCreate(savedInstanceState);
    setContentView(R.layout.activity_main);

    TextView tv = (TextView) findViewById(R.id.id_tv);
    Animation rotate = AnimationUtils.loadAnimation(this,
R.anim.my_rotate);
    rotate.setAnimationListener(new AnimationListener() {

        @Override
        public void onAnimationStart(Animation animation) {
        }

        @Override
        public void onAnimationEnd(Animation animation) {
            Toast.makeText(MainActivity.this, "动画结束",
                            Toast.LENGTH_LONG).show();
        }

        @Override
        public void onAnimationRepeat(Animation animation) {
        }

    });
```

```
        tv.startAnimation(rotate);
    }
```

在修改后的 onCreate()回调方法中，通过调用 rotate.setAnimationListener()方法设置旋转动画的监听器，并且在 onAnimationEnd()方法中显示结束消息。运行修改后的 Ch0901 应用程序，旋转动画执行完毕后的界面如图 9-4 所示。

图 9-4　旋转动画执行完毕后的界面

9.2　View 动画之帧动画

帧动画的基本原理是按一定的时间顺序显示一组预定义的图片。在使用 XML 文件定义帧动画时，需要将帧动画定义文件存储于工程的 res/drawable 目录下，文件名可以是任意合法的文件名。在 XML 文件或在 Java 类文件中引用帧动画资源的形式与引用一般图片资源的形式是一样的，可以在任何可以使用 drawable 资源的地方使用帧动画资源。定义帧动画的一般格式如下：

```
<?xml version="1.0" encoding="utf-8"?>
<animation-list
    xmlns:android="http://schemas.android.com/apk/res/android"
    android:oneshot=["true" | "false"] >
    <item
        android:drawable="@[package:]drawable/drawable_resource_name"
                                    android:duration="integer" />
    <item …/>
```

```
        …
    </animation-list>
```

以下代码主要用于指定是循环执行动画,还是一次性执行动画。当值为 true 时,表示一次性执行动画;当值为 false 时,表示循环执行动画。

```
    android:oneshot=["true" | "false"]>
```

以下代码主要用于定义在帧动画中要显示的图片,android:drawable 属性主要用于指定要显示的图片,android:duration 属性主要用于指定图片显示的时间。

```
    <item
        android:drawable="@[package:]drawable/drawable_resource_name"
                                        android:duration="integer" />
    <item …/>
        …
```

下面举例说明如何在 Android 应用程序中使用帧动画:按顺序显示一组图片,如图 9-5 所示,每张图片显示 200 毫秒。

图 9-5　帧动画案例要显示的一组图片

该案例应用程序的运行效果如图 9-6 所示。

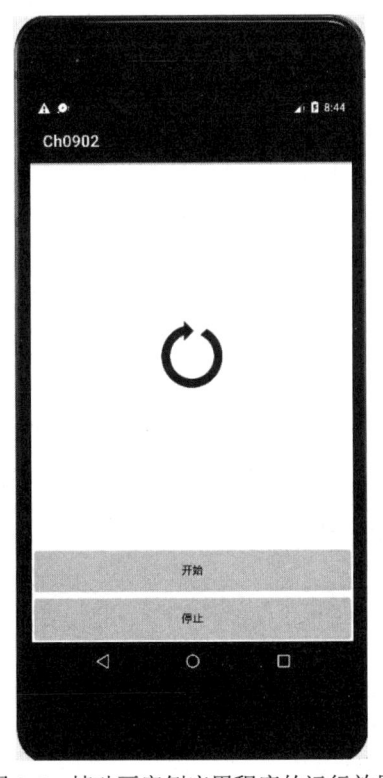

图 9-6　帧动画案例应用程序的运行效果

下面构建该案例应用程序。在 Android Studio 中新建一个名为 Ch0902 的 Android 应用程序工程。在 res/drawable 目录下存储图 9-5 所示的图片,并且新建一个名为 drawable_rotate.xml 的帧动画资源文件。drawable_rotate.xml 文件中的代码如下:

```xml
<?xml version="1.0" encoding="utf-8"?>
<animation-list xmlns:android="http://schemas.android.com/apk/res/android"
    android:oneshot="false">
    <item android:drawable="@drawable/z01" android:duration="200" />
    <item android:drawable="@drawable/z02" android:duration="200" />
    <item android:drawable="@drawable/z03" android:duration="200" />
    <item android:drawable="@drawable/z04" android:duration="200" />
</animation-list>
```

在 drawable_rotate.xml 文件中,指定 Ch0902 应用程序按顺序显示 4 张图片,并且设置每张图片的显示时间为 200 毫秒。

修改布局文件 res/layout/activity_main.xml 中的代码,在 LinearLayout 容器组件中放置一个 ImageView 组件和两个 Button 组件,修改后的代码如下:

```xml
<LinearLayout xmlns:android="http://schemas.android.com/apk/res/android"
    android:layout_width="match_parent"
    android:layout_height="match_parent"
    android:orientation="vertical">

    <android.widget.ImageView
        android:id="@+id/id_iamge_view"
        android:layout_width="match_parent"
        android:layout_height="0dp"
        android:layout_weight="8"
        android:scaleType="center"
        android:src="@drawable/drawable_rotate"
        android:contentDescription="@string/hello_world"/>

    <android.widget.Button
        android:id="@+id/id_button_1"
        android:layout_width="match_parent"
        android:layout_height="0dp"
        android:layout_weight="1"
        android:text="@string/text_start"
        />

    <android.widget.Button
        android:id="@+id/id_button_2"
        android:layout_width="match_parent"
        android:layout_height="0dp"
        android:layout_weight="1"
```

```xml
        android:text="@string/text_stop"
        />
```

```xml
</LinearLayout>
```

修改 res/values/strings.xml 文件中的代码，定义布局文件中引用的字符串资源，修改后的代码如下：

```xml
<resources>
    <string name="app_name">Ch0902</string>

    <string name="hello_world">Hello world!</string>
    <string name="action_settings">Settings</string>
    <string name="text_start">开始</string>
    <string name="text_stop">停止</string>

</resources>
```

修改 MainActivity.java 文件中的代码，在点击"开始"按钮后，开始执行动画；在点击"停止"按钮后，停止执行动画，修改后的代码如下：

```java
package com.example.ch0902;

import androidx.appcompat.app.AppCompatActivity;

import android.graphics.drawable.AnimationDrawable;
import android.os.Bundle;
import android.view.View;
import android.widget.Button;
import android.widget.ImageView;

public class MainActivity extends AppCompatActivity implements View.OnClickListener{

    @Override
    protected void onCreate(Bundle savedInstanceState) {
        super.onCreate(savedInstanceState);
        setContentView(R.layout.activity_main);
        Button btn1 = (Button) this.findViewById(R.id.id_button_1);
        btn1.setOnClickListener(this);
        Button btn2 = (Button) this.findViewById(R.id.id_button_2);
        btn2.setOnClickListener(this);
    }

    @Override
    public void onClick(View v) {
```

```
        int id = v.getId();
        if (id == R.id.id_button_1) {
            ImageView iv = (ImageView)this.findViewById(R.id.id_iamge_view);
            AnimationDrawable ad = (AnimationDrawable)iv.getDrawable();
            ad.start();
        }
        else {
            ImageView iv = (ImageView)this.findViewById(R.id.id_iamge_view);
            AnimationDrawable ad = (AnimationDrawable)iv.getDrawable();
            ad.stop();
        }
    }
}
```

在对按钮点击事件的响应方法 onClick()中，从界面中获取 ImageView 组件的引用，并且获取在其中显示的 Drawable 对象，由于 ImageView 组件的 Drawable 对象是一个帧动画，因此在获取该帧动画资源后，可以通过 start()方法或 stop()方法开始或停止执行动画。

执行 Ch0902 应用程序，即可得到图 9-6 所示的运行效果，点击"开始"或"停止"按钮，可以开始或停止执行动画。

9.3 同步练习

编写一个简单的动画应用程序，在应用程序启动后，在一个 ImageView 组件中执行一个开启动画；在动画执行完毕后，显示应用程序的主界面。

第10章

多媒体播放

Android 的多媒体框架提供了播放音频、视频及图像的相关方法。通过该框架，我们可以处理来自程序资源、本地文件系统及网络中的多媒体内容。

MediaPlayer 是 Android 多媒体框架中的重要类。通过使用 MediaPlayer 对象，Android 应用程序可以获取、解码和播放音频、视频等多媒体资源。播放的多媒体资源可以来自程序资源（存储于 res/raw 目录下，因为 Android 不会对存储于该目录下的资源进行任何处理，所以存储于该目录下的文件可以保持原样）、本地文件系统和网络。

本章主要介绍使用 MediaPlayer 对象播放音频和视频的相关知识。

10.1 播放音频

在 Android 应用程序中播放的音频大致可以分为两类：为 Android 应用程序增加音效的简短音频和来自本地文件系统或网络的音频。对于为 Android 应用程序增加音效的简短音频，可以使用 MediaPlayer 提供的便利播放方式进行播放；对于来自本地文件系统或网络的音频，编者建议使用 MediaPlayer 提供的完整播放方式进行播放。

10.1.1 播放简短音频

为了播放简短音频，MediaPlayer 提供了几种静态的 create()方法，用于创建 MediaPlayer 对象。查看 Android 帮助文档可知，MediaPlayer 提供的便利的 create()方法如图 10-1 所示。

static MediaPlayer	create (Context context, int resid) Convenience method to create a MediaPlayer for a given resource id.
static MediaPlayer	create (Context context, Uri uri) Convenience method to create a MediaPlayer for a given Uri.

图 10-1 MediaPlayer 提供的便利的 create()方法

其中，create(Context context, int resid)方法主要用于播放存储于 res/raw 目录下的音频资源，create(Context context, Uri uri)方法主要用于播放来自任何 Uri 的音频资源。对于需要播放的简短音频，编者建议将音频资源存储于 res/raw 目录下。

下面举例说明如何使用这种方式播放简短音频。在该案例应用程序运行后，会执行一个时间长度为 4 秒的启动动画，并且同步循环播放一段音频（因为音频的时间长度少于 4 秒，所以需要循环播放），在启动动画执行完毕时，停止播放简短音频。该案例应用程序的运行效果如图 10-2 所示。

图 10-2　案例应用程序的运行效果

下面构建该案例应用程序。在 Android Studio 中新建一个名为 Ch1001 的 Android 应用程序工程，并且在工程的 res 目录下分别新建名为 raw 和 anim 的子目录。在 res/raw 目录下存储一个要播放的音频文件，其文件名为 ring08.wav。修改布局文件 res/layout/activity_main.xml 中的代码，修改后的代码如下：

```xml
<RelativeLayout xmlns:android="http://schemas.android.com/apk/res/android"
    android:layout_width="match_parent"
    android:layout_height="match_parent">

    <ImageView
        android:id="@+id/id_image_view"
        android:layout_width="match_parent"
        android:layout_height="match_parent"
        android:scaleType="fitCenter"
        android:src="@mipmap/ic_launcher"
        android:contentDescription="hello_world" />

</RelativeLayout>
```

这个布局文件很简单，只包含一个 ImageView 组件。

在 res/anim 目录下新建一个名为 my_scale.xml 的动画定义文件，该文件中的代码如下：

```xml
<?xml version="1.0" encoding="utf-8"?>
<scale xmlns:android="http://schemas.android.com/apk/res/android"
```

```
        android:interpolator="@android:anim/linear_interpolator"
        android:fromXScale="0"
        android:toXScale="1"
        android:fromYScale="0"
        android:toYScale="1"
        android:pivotX="50%"
        android:pivotY="50%"
        android:duration="4000"
        android:fillAfter="true"
        >
</scale>
```

这个动画文件会以组件中心点为固定点，在 4 秒内，将组件在 x 轴方向和 y 轴方向上从 0.0 放大到组件的正常尺寸。

修改 MainActivity.java 文件中的代码，修改后的代码如下：

```
package com.example.ch1001;

import androidx.appcompat.app.AppCompatActivity;

import android.media.MediaPlayer;
import android.os.Bundle;
import android.view.animation.Animation;
import android.view.animation.AnimationUtils;
import android.widget.ImageView;

public class MainActivity extends AppCompatActivity {
    private MediaPlayer mp;

    @Override
    protected void onCreate(Bundle savedInstanceState) {
        super.onCreate(savedInstanceState);
        setContentView(R.layout.activity_main);

        mp = null;

        ImageView iv = (ImageView) findViewById(R.id.id_image_view);
        Animation rotate = AnimationUtils.loadAnimation(this, R.anim.my_scale);
        rotate.setAnimationListener(new Animation.AnimationListener() {

            @Override
            public void onAnimationStart(Animation animation) {
                mp = MediaPlayer.create(MainActivity.this, R.raw.ring08);
                mp.setLooping(true);
```

```
            mp.setVolume(0.5f, 0.5f);
            mp.start();
        }

        @Override
        public void onAnimationEnd(Animation animation) {
            mp.stop();
            mp.release();
            mp = null;
        }

        @Override
        public void onAnimationRepeat(Animation animation) {

        }

    });
    iv.startAnimation(rotate);
}
```

主要工作都是在 MainActivity 类的 onCreate()方法中完成的。在该方法中，首先获取 ImageView 组件的引用，然后加载动画资源，并且设置动画监听器。在动画监听器的 onAnimationStart()方法中，通过以下代码启动并播放一个简短音频。

```
mp = MediaPlayer.create(MainActivity.this, R.raw.short4);
mp.setLooping(true);
mp.setVolume(0.5f, 0.5f);
mp.start();
```

在上述代码中，使用 MediaPlayer.create()方法创建了 MediaPlayer 对象 mp，同时加载并解码了本地音频资源。需要注意的是，对于来自网络的音频资源，MediaPlayer.create()方法可能需要较长的时间进行加载和解码,因此,当在 Android 生成的 UI 线程中执行 MediaPlayer.create()方法时，可能出现"响应超时"错误。在加载并解码了音频资源后，首先调用 mp 对象的 setLooping()方法，用于设置循环播放该音频；然后调用 mp 对象的 setVolume()方法，用于设置该音频的播放音量；最后调用 mp 对象的 start()方法，用于播放该音频。在动画监听器的 onAnimationEnd()方法中，通过以下代码停止播放音频，并且释放 mp 对象占用的系统资源。需要注意的是，在停止播放并关闭播放器时，一定要调用 mp 对象的 release()方法，用于释放其占用的资源。

```
mp.stop();
mp.release();
mp = null;
```

运行 Ch1001 应用程序，即可得到图 10-2 所示的运行效果，在 ImageView 组件逐渐放大的同时会播放一段音频。

10.1.2 使用 MediaPlayer 自制一个音频播放器

下面举例说明如何编写一个播放本地 SD 卡中音频文件的播放器应用程序，用于介绍如何使用 MediaPlayer 提供的完整播放方式播放音频。该案例应用程序首先显示 SD 卡中的目录及文件列表，运行效果如图 10-3 所示。此时，点击任意一个文件，该应用程序都会播放所点击的文件，并且在列表右边显示一个播放图标，用于表示正在播放此文件，同时，在界面上方的"正在播放"文本框中显示正在播放的文件。我们还可以点击列表中的目录名称进入相应的目录，点击界面左上角的箭头可以返回上一级目录。

下面构建该案例应用程序。在 Android Studio 中新建一个名为 Ch1002 的 Android 应用程序工程，然后在 res/mipmap 目录下存储需要用到的图标，包括文件浏览图标（界面左上角的箭头）、文件类型图标和播放图标。

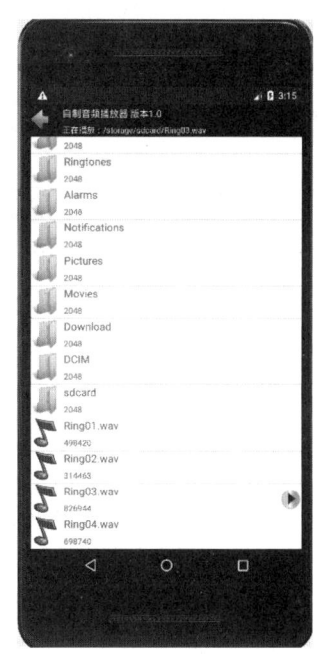

图 10-3 自制音频播放器案例应用程序的运行效果

修改主界面布局文件 res/layout/activity_main.xml 中的代码，修改后的代码如下：

```xml
<LinearLayout xmlns:android="http://schemas.android.com/apk/res/android"
    android:layout_width="match_parent"
    android:layout_height="match_parent"
    android:orientation="vertical">

    <LinearLayout
        android:layout_width="match_parent"
        android:layout_height="48dp"
        android:background="#FF000000"
        android:orientation="horizontal" >

        <ImageButton
            android:id="@+id/id_image_button"
            android:layout_width="32dp"
            android:layout_height="32dp"
            android:layout_gravity="center"
            android:background="@mipmap/png1638"
            android:scaleType="fitCenter"
            android:contentDescription="@string/hello_world"
        />

        <View
            android:layout_width="16dp"
```

```xml
        android:layout_height="match_parent"
        />

    <LinearLayout
        android:layout_width="match_parent"
        android:layout_height="match_parent"
        android:orientation="vertical"
        >

        <TextView
            android:id="@+id/id_text_view_1"
            android:layout_width="match_parent"
            android:layout_height="28dp"
            android:gravity="center_vertical"
            android:textSize="14sp"
            android:textColor="#FFFFFFFF"
            android:text="@string/text_version"
        />

        <TextView
            android:id="@+id/id_text_view_2"
            android:layout_width="match_parent"
            android:layout_height="20dp"
            android:gravity="center_vertical"
            android:textSize="12sp"
            android:textColor="#FFFFFFFF"
            android:text="@string/text_current_playing"
        />

    </LinearLayout>

</LinearLayout>

<ListView
    android:id="@+id/id_list_view"
    android:layout_width="match_parent"
    android:layout_height="match_parent"
    />

</LinearLayout>
```

activity_main.xml 文件主要用于设置图 10-3 所示的界面布局，该文件中包含一个 ListView 组件，在 ListView 组件中显示目录浏览图标、软件名称和版本、正在播放的文件。由于在该布局文件中引用了一些字符串资源，因此需要修改 res/values/strings.xml 文件中的代码，修改后的代码如下：

```xml
<resources>
    <string name="app_name">Ch1002</string>

    <string name="hello_world">Hello world!</string>
    <string name="text_version">自制音频播放器 版本 1.0</string>
    <string name="text_current_playing">正在播放：</string>

</resources>
```

创建 ListView 组件列表项的布局文件 res/layout/list_item.xml，然后修改该文件中的代码，修改后的代码如下：

```xml
<?xml version="1.0" encoding="utf-8"?>
<RelativeLayout xmlns:android="http://schemas.android.com/apk/res/android"
    android:layout_width="match_parent"
    android:layout_height="48dp"
    android:layout_gravity="center_vertical"
    >

    <ImageView
        android:id="@+id/id_li_image_view_type"
        android:layout_width="44dp"
        android:layout_height="44dp"
        android:layout_alignParentLeft="true"
        android:scaleType="fitCenter"
        android:contentDescription="@string/hello_world"
        />

    <View
        android:id="@+id/id_li_space_1"
        android:layout_width="6dp"
        android:layout_height="44dp"
        android:layout_toRightOf="@id/id_li_image_view_type"
        />

    <LinearLayout
        android:layout_width="wrap_content"
        android:layout_height="44dp"
        android:layout_toRightOf="@id/id_li_space_1"
        android:orientation="vertical" >

        <TextView
```

```xml
            android:id="@+id/id_li_text_view_name"
            android:layout_width="wrap_content"
            android:layout_height="28dp"
            android:textSize="16sp"
            />

        <TextView
            android:id="@+id/id_li_text_view_length"
            android:layout_width="wrap_content"
            android:layout_height="16dp"
            android:textSize="12sp"
            />

    </LinearLayout>

    <ImageView
        android:id="@+id/id_li_image_view_play"
        android:layout_width="32dp"
        android:layout_height="32dp"
        android:layout_alignParentRight="true"
        android:layout_centerVertical="true"
        android:scaleType="fitCenter"
        android:contentDescription="@string/hello_world"
        />

</RelativeLayout>
```

在列表项中显示文件类型图标、文件名、文件长度及播放图标，如图 10-4 所示。

图 10-4 列表项布局示例

为了在 ListView 组件中显示文件列表，需要构建 Adapter 接口的一个实现类，用于为 ListView 组件提供数据。因此，在工程的 src 目录下，新建一个名为 com.example.ch1002.adapter 的包，并且在该包中新建一个名为 FileListAdapter 的 Java 类文件，该文件中的代码如下：

```java
package com.example.ch1002.adapter;

import android.content.Context;
import android.os.Environment;
import android.view.LayoutInflater;
import android.view.View;
import android.view.ViewGroup;
import android.widget.BaseAdapter;
```

```java
import android.widget.ImageView;
import android.widget.ListView;
import android.widget.RelativeLayout;
import android.widget.TextView;

import java.io.File;
import java.util.ArrayList;

import com.example.ch1002.R;

public class FileListAdapter extends BaseAdapter {
    private File cFile;
    private ArrayList<FileDesc> cList;

    private String currentPlaying;

    LayoutInflater inflater;

    public FileListAdapter(Context context) {
        inflater = (LayoutInflater)
                context.getSystemService(Context.LAYOUT_INFLATER_SERVICE);
        currentPlaying = null;

        cList = new ArrayList<FileDesc>();
        cFile = Environment.getExternalStorageDirectory();

        load();
    }

    @Override
    public int getCount() {
        return cList.size();
    }

    @Override
    public Object getItem(int position) {
        return null;
    }

    @Override
    public long getItemId(int position) {
        return 0;
    }
```

```java
    @Override
    public View getView(int position, View convertView, ViewGroup parent) {
        FileDesc fd = cList.get(position);
        RelativeLayout rl;

        if (convertView != null)
            rl = (RelativeLayout)convertView;
        else
            rl = (RelativeLayout)inflater.inflate(R.layout.list_item, parent, false);

        rl.setTag(fd.file.getAbsolutePath());

        ImageView iv_type = (ImageView)rl.findViewById(R.id.id_li_image_view_type);
        if (fd.type.equalsIgnoreCase("dir"))
            iv_type.setImageResource(R.mipmap.png1674);
        else
            iv_type.setImageResource(R.mipmap.png0088);

        TextView tv_name = (TextView)rl.findViewById(R.id.id_li_text_view_name);
        tv_name.setText(fd.name);

        TextView tv_length = (TextView)rl.findViewById(R.id.id_li_text_view_length);
        tv_length.setText(fd.length + "");

        ImageView iv_play = (ImageView)rl.findViewById(R.id.id_li_image_view_play);
        iv_play.setImageResource(R.mipmap.png1702);
        if ((fd.file.getAbsolutePath()).equalsIgnoreCase(currentPlaying))
            iv_play.setVisibility(View.VISIBLE);
        else
            iv_play.setVisibility(View.INVISIBLE);

        return rl;
    }

    private void load() {
        File[] fl = cFile.listFiles();
        if (fl == null)
            return;

        cList.clear();

        int count = fl.length;
```

```java
        for(int i=0; i<count; i++) {
            FileDesc fd = new FileDesc();
            fd.file = fl[i];
            fd.name = fl[i].getName();
            fd.length = fl[i].length();
            fd.type = fl[i].isDirectory()==true?"dir":"file";
            
            cList.add(fd);
        }
        
        this.notifyDataSetChanged();
    }
    
    public void goUp() {
        if (Environment.getExternalStorageDirectory().getAbsolutePath().
                equalsIgnoreCase(cFile.getAbsolutePath())) {
            return;
        }
        
        cFile = cFile.getParentFile();
        load();
    }
    
    public void goDown(File ndir) {
        cFile = ndir;
        load();
    }
    
    public String getType(int position) {
        return cList.get(position).type;
    }
    
    public File getFile(int position) {
        return cList.get(position).file;
    }
    
    public File setAndPlay(ListView lv, int position) {
        FileDesc fd = cList.get(position);
        RelativeLayout rl;
        
        if (currentPlaying != null) {
            rl = (RelativeLayout)lv.findViewWithTag(currentPlaying);
            if (rl != null) {
```

```
            ImageView iv_play =
                    (ImageView)rl.findViewById(R.id.id_li_image_view_play);
            iv_play.setVisibility(View.INVISIBLE);
        }
    }

    currentPlaying = fd.file.getAbsolutePath();
    rl = (RelativeLayout)lv.findViewWithTag(currentPlaying);
    if (rl != null) {
        ImageView iv_play = (ImageView)rl.findViewById(R.id.id_li_image_view_play);
        iv_play.setVisibility(View.VISIBLE);
    }

    return cList.get(position).file;
}

private class FileDesc {
    public File file;
    public String name;
    public long length;
    public String type;
}
```

下面来看其中的变量定义，代码如下：

```
private File cFile;
private ArrayList<FileDesc> cList;

private String currentPlaying;
```

其中的 cFile 表示当前主界面中显示的目录 File 对象，cList 表示当前目录的文件列表。FileDesc 是一个自定义的类，用于表示要展示的文件信息，FileDesc 类的定义代码如下：

```
private class FileDesc {
    public File file;
    public String name;
    public long length;
    public String type;
}
```

我们只需要展示文件的文件名、长度和类型，为了便于后续操作，将每个文件的 File 对象存储于 FileDesc 对象中。

FileListAdapter 类的构造方法如下：

```
public FileListAdapter(Context context) {
    inflater = (LayoutInflater)
            context.getSystemService(Context.LAYOUT_INFLATER_SERVICE);
    currentPlaying = null;

    cList = new ArrayList<FileDesc>();
    cFile = Environment.getExternalStorageDirectory();

    load();
}
```

在该构造方法中会创建一个 ArrayList 对象 cList，通过调用 Environment 类中的静态方法 getExternalStorageDirectory()获取 SD 卡的首目录，并且将 cFile 对象表示的目录设置为 SD 卡的首目录。Environment 类很强大，可以获取很多与文件系统有关的信息，其中常用的方法如图 10-5 所示。

Public Methods	
static File	getDataDirectory() Return the user data directory.
static File	getDownloadCacheDirectory() Return the download/cache content directory.
static File	getExternalStorageDirectory() Return the primary external storage directory.
static File	getExternalStoragePublicDirectory(String type) Get a top-level public external storage directory for placing files of a particular type.
static String	getExternalStorageState(File path) Returns the current state of the storage device that provides the given path.
static String	getExternalStorageState() Returns the current state of the primary "external" storage device.
static File	getRootDirectory() Return root of the "system" partition holding the core Android OS.

图 10-5 Environment 类中常用的方法

通过调用 load()方法加载当前的目录信息。load()方法的定义代码如下：

```
private void load() {
    File[] fl = cFile.listFiles();
    if (fl == null)
        return;

    cList.clear();

    int count = fl.length;
    for(int i=0; i<count; i++) {
        FileDesc fd = new FileDesc();
        fd.file = fl[i];
        fd.name = fl[i].getName();
        fd.length = fl[i].length();
        fd.type = fl[i].isDirectory()==true?"dir":"file";
```

```
                cList.add(fd);
        }

        this.notifyDataSetChanged();
    }
```

在 load()方法中，首先获取当前目录的文件信息列表，然后对其进行逐项处理，将其存储于 cList 对象中，最后通知 ListView 组件显示列表信息。

FileListAdapter 类中的其他方法包括 getCount()、getItem()、getItemId()和 getView()，下面详细介绍 getView()方法，该方法的定义代码如下：

```
    public View getView(int position, View convertView, ViewGroup parent) {
        FileDesc fd = cList.get(position);
        RelativeLayout rl;

        if (convertView != null)
            rl = (RelativeLayout)convertView;
        else
            rl = (RelativeLayout)inflater.inflate(R.layout.list_item, parent, false);

        rl.setTag(fd.file.getAbsolutePath());

        ImageView iv_type = (ImageView)rl.findViewById(R.id.id_li_image_view_type);
        if (fd.type.equalsIgnoreCase("dir"))
            iv_type.setImageResource(R.drawable.png1674);
        else
            iv_type.setImageResource(R.drawable.png0088);

        TextView tv_name = (TextView)rl.findViewById(R.id.id_li_text_view_name);
        tv_name.setText(fd.name);

        TextView tv_length = (TextView)rl.findViewById(R.id.id_li_text_view_length);
        tv_length.setText(fd.length + "");

        ImageView iv_play = (ImageView)rl.findViewById(R.id.id_li_image_view_play);
        iv_play.setImageResource(R.drawable.png1702);
        if ((fd.file.getAbsolutePath()).equalsIgnoreCase(currentPlaying))
            iv_play.setVisibility(View.VISIBLE);
        else
            iv_play.setVisibility(View.INVISIBLE);

        return rl;
    }
```

在 getView()方法中，首先重用或新建一个 RelativeLayout 对象 rl，用于表示列表项；然后通过以下代码为每个列表项都设置一个 Tag，用于在后续操作中，使用 findViewWithTag()方法获取所需的列表项；最后设置列表项应该显示的信息，包括文件类型图标、文件名、文件长度，如果当前正在播放某个文件，则需要在该文件的列表项中显示播放图标。

```
rl.setTag(fd.file.getAbsolutePath());
```

FileListAdapter 类中还包括 goUP()方法、goDown()方法、getType()方法、getFile()方法及 setAndPlay()方法，这些方法会被 MainActivity 对象调用，分别用于显示上一级目录、显示下一级目录、得到指定位置的列表项所显示的文件类型（目录或普通文件）、得到指定位置的列表项的 File 对象及设置播放图标。下面对 setAndPlay()方法进行简单的介绍，该方法的定义代码如下：

```java
    public File setAndPlay(ListView lv, int position) {
        FileDesc fd = cList.get(position);
        RelativeLayout rl;

        if (currentPlaying != null) {
            rl = (RelativeLayout)lv.findViewWithTag(currentPlaying);
            if (rl != null) {
                ImageView iv_play =
                    (ImageView)rl.findViewById(R.id.id_li_image_view_play);
                iv_play.setVisibility(View.INVISIBLE);
            }
        }

        currentPlaying = fd.file.getAbsolutePath();
        rl = (RelativeLayout)lv.findViewWithTag(currentPlaying);
        if (rl != null) {
            ImageView iv_play = (ImageView)rl.findViewById(R.id.id_li_image_view_play);
            iv_play.setVisibility(View.VISIBLE);
        }

        return cList.get(position).file;
    }
```

setAndPlay()方法主要用于在被播放文件的列表项中显示播放图标，其实现方式如下：判断当前是否存在正在播放的文件，如果存在正在播放的文件，则隐藏当前列表项的播放图标，并且在即将播放的文件的列表项中显示播放图标。

修改 MainActivity.java 文件中的代码，修改后的代码如下：

```java
package com.example.ch1002;

import androidx.appcompat.app.ActionBar;
import androidx.appcompat.app.AppCompatActivity;
```

```java
import android.media.AudioManager;
import android.media.MediaPlayer;
import android.os.Bundle;
import android.view.View;
import android.widget.AdapterView;
import android.widget.ImageButton;
import android.widget.ListView;
import android.widget.TextView;
import android.widget.Toast;

import com.example.ch1002.adapter.FileListAdapter;

import java.io.File;
import java.io.IOException;

public class MainActivity extends AppCompatActivity
        implements AdapterView.OnItemClickListener, View.OnClickListener{
    private ListView lv;
    private ImageButton ib;
    private TextView tv;

    private FileListAdapter fla;

    private MediaPlayer mediaPlayer;

    @Override
    protected void onCreate(Bundle savedInstanceState) {
        super.onCreate(savedInstanceState);
        setContentView(R.layout.activity_main);

        ActionBar ab = this.getSupportActionBar();
        if (ab != null) ab.hide();

        lv = (ListView)this.findViewById(R.id.id_list_view);
        ib = (ImageButton)this.findViewById(R.id.id_image_button);
        tv = (TextView)this.findViewById(R.id.id_text_view_2);

        fla = new FileListAdapter(this);
        lv.setAdapter(fla);
        lv.setOnItemClickListener(this);

        ib.setOnClickListener(this);
```

```java
        }

        @Override
        public void onItemClick(AdapterView<?> parent, View view, int position, long id) {
            String type = fla.getType(position);
            if (type.equalsIgnoreCase("dir")) {
                File ndir = fla.getFile(position);
                fla.goDown(ndir);
            }
            else {
                File nf = fla.setAndPlay(lv, position);
                tv.setText("正在播放: " + nf.getAbsolutePath());
                play(nf);
            }
        }

        @Override
        public void onClick(View v) {
            int id = v.getId();
            if (id == R.id.id_image_button) {
                fla.goUp();
            }
        }

        private void play(File file) {
            if (mediaPlayer != null) {
                mediaPlayer.stop();
                mediaPlayer.release();
                mediaPlayer = null;
            }

            mediaPlayer = new MediaPlayer();
            mediaPlayer.setAudioStreamType(AudioManager.STREAM_MUSIC);
            mediaPlayer.setOnPreparedListener(new MediaPlayer.OnPreparedListener() {
                @Override
                public void onPrepared(MediaPlayer mp) {
                    mediaPlayer.start();
                }
            });
            mediaPlayer.setOnErrorListener(new MediaPlayer.OnErrorListener() {
                @Override
                public boolean onError(MediaPlayer mp, int what, int extra) {
```

```
                Toast.makeText(MainActivity.this, "无法播放该文件",
                        Toast.LENGTH_LONG).show();
                return false;
            }
        });

        try {
            mediaPlayer.setDataSource(file.getAbsolutePath());
            mediaPlayer.prepareAsync();
        } catch (IllegalArgumentException e) {
            e.printStackTrace();
        } catch (SecurityException e) {
            e.printStackTrace();
        } catch (IllegalStateException e) {
            e.printStackTrace();
        } catch (IOException e) {
            e.printStackTrace();
        }
    }
}
```

在 MainActivity.java 文件中，onCreate()回调方法主要用于获取界面中各个组件的引用，并且设置相应的监听接口。

对列表项点击事件的响应过程是在 onItemClick()方法中完成的，代码如下：

```
    public void onItemClick(AdapterView<?> parent, View view, int position,
long id) {
        String type = fla.getType(position);
        if (type.equalsIgnoreCase("dir")) {
            File ndir = fla.getFile(position);
            fla.goDown(ndir);
        }
        else {
            File nf = fla.setAndPlay(lv, position);
            tv.setText("正在播放: " + nf.getAbsolutePath());
            play(nf);
        }
    }
```

在 onItemClick()方法中，应用程序会判断所点击的列表项是目录还是普通文件，如果是目录，则使用 FileListAdapter 对象调用 goDown()方法，在 ListView 组件中显示下一级目录的内容；如果是普通文件，则首先显示播放图标，然后在主界面的上方显示正在播放文件的文件名，最后调用 play()方法播放所点击的文件。play()方法的定义代码如下：

```
    private void play(File file) {
        if (mediaPlayer != null) {
```

```
            mediaPlayer.stop();
            mediaPlayer.release();
            mediaPlayer = null;
        }

        mediaPlayer = new MediaPlayer();
        mediaPlayer.setAudioStreamType(AudioManager.STREAM_MUSIC);
        mediaPlayer.setOnPreparedListener(new OnPreparedListener() {
            @Override
            public void onPrepared(MediaPlayer mp) {
                mediaPlayer.start();
            }
        });
        mediaPlayer.setOnErrorListener(new OnErrorListener() {
            @Override
            public boolean onError(MediaPlayer mp, int what, int extra) {
                Toast.makeText(MainActivity.this, "无法播放该文件",
                               Toast.LENGTH_LONG).show();
                return false;
            }
        });

        try {
            mediaPlayer.setDataSource(file.getAbsolutePath());
            mediaPlayer.prepareAsync();
        } catch (IllegalArgumentException e) {
            e.printStackTrace();
        } catch (SecurityException e) {
            e.printStackTrace();
        } catch (IllegalStateException e) {
            e.printStackTrace();
        } catch (IOException e) {
            e.printStackTrace();
        }
    }
```

在 play()方法中，先判断当前是否正在播放某个文件，若是，则停止播放该文件；再新建一个 MediaPlayer 对象 mediaPlayer，设置其可以播放音频文件，并且设置 OnPreparedListener 接口。现在的问题是，为什么要这么做呢？难道不能使用在 10.1.1 节中介绍的简单方法吗？回答是不能。读者可以思考这个问题：假设要播放来自网络的某个音频文件，但网络的速率是不可预期的，那么加载要播放的音频文件可能需要很长的时间，从而导致 Activity 响应超时。因此，对于大的或来自网络的音频文件，需要调用 setOnPreparedListener()方法，使用异步加载的方式播放音频文件。OnPreparedListener 接口中只有一个方法：onPrepared()。onPrepared()方法表示对音频文件的加载和解码已经完成，可以播放了。因此，可以在 onPrepared()方法中直接播放音频文件。

对于 mediaPlayer 对象，我们还为其设置了 OnErrorListener 接口，用于监听在加载、解码及播放过程中出现的任何错误（只是显示一个错误消息而已）。

在设置了相关接口后，调用 setDataSource()方法设置 mediaPlayer 对象即将播放的文件，并且调用 prepareAsync()方法进行异步加载和解码，代码如下：

```
mediaPlayer.setDataSource(file.getAbsolutePath());
mediaPlayer.prepareAsync();
```

为了能够播放音频文件，还需要上传一些音频文件到模拟器中。

在 Android Studio 中单击最右边的 Device File Explorer 按钮，如图 10-6 所示，即可打开 Device File Explorer 面板，用于向模拟器中上传音频文件，如图 10-7 所示。

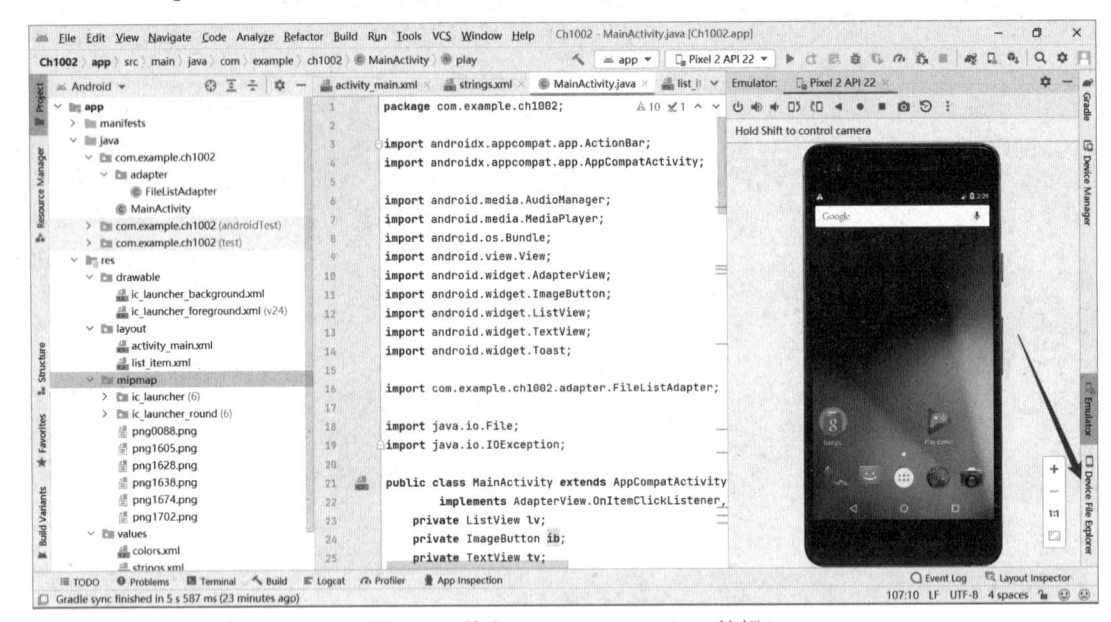

图 10-6　单击 Device File Explorer 按钮

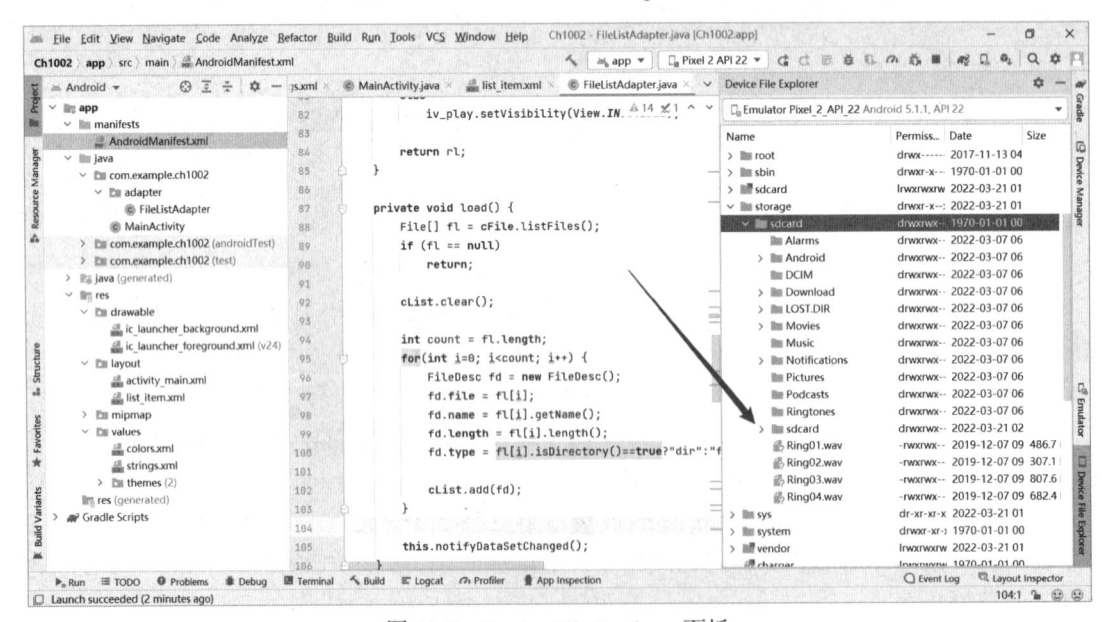

图 10-7　Device File Explorer 面板

右击 sdcard→sdcard 选项，在弹出的快捷菜单中选择 Upload 命令，然后在计算机中选择要上传到 sdcard 目录下的文件，即可将计算机中的文件上传到模拟器中。

在 AndroidManifest.xml 文件中为应用程序访问 SD 卡授权，代码如下：

```
<uses-permission android:name="android.permission.READ_EXTERNAL_STORAGE"/>
```

运行 Ch1002 应用程序，即可得到图 10-3 所示的运行效果。

10.2 同步练习一

将 10.1.2 节中的 Ch1002 应用程序加载到 Android Studio 中并运行，然后进行以下修改。
- 使界面更加美观。
- 在列表中只显示能够播放的音频文件和子目录。
- 点击手机上的返回键图标，显示当前目录的上一级目录。

10.3 播放视频

我们可以使用 MediaPlayer 组件播放视频。要播放视频，需要先创建一个用于显示视频的组件。在此，我们使用 Android 的 VideoView 组件播放视频。下面举例说明如何使用 VideoView 组件播放视频。

在播放视频前，先使用 10.1.2 节中介绍的方法将一个 MP4 视频文件（bob.mp4）上传到模拟器的 sdcard 目录下，如图 10-8 所示。

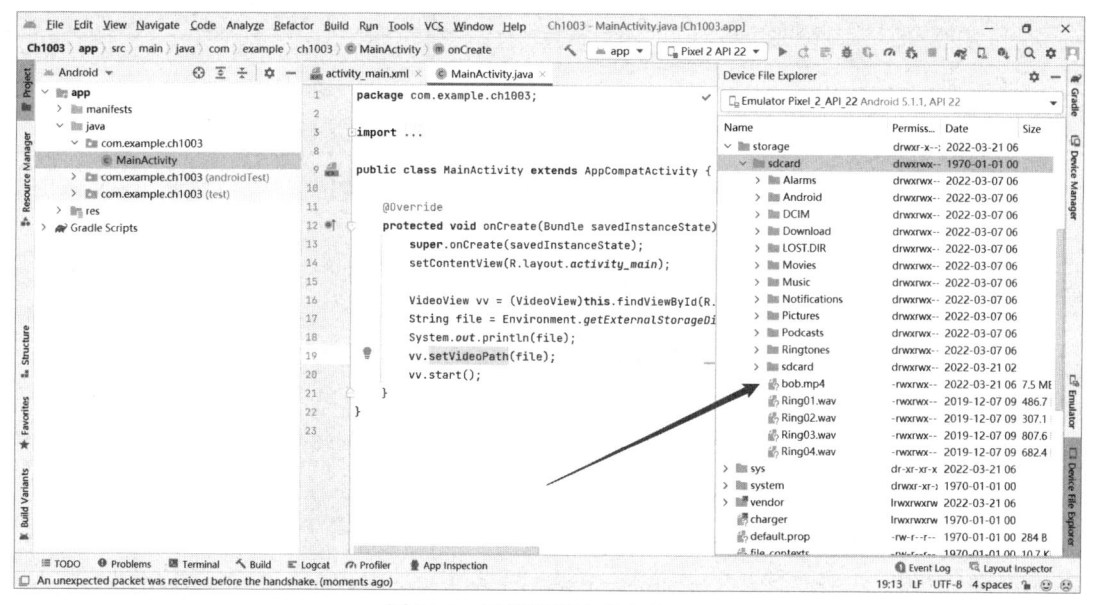

图 10-8 示例视频文件 bob.mp4

在文件上传完成后，运行该案例应用程序，即可播放该示例视频文件 bob.mp4，运行效果如图 10-9 所示（Android 手机模拟器可能不支持某些视频格式，为了测试该案例应用程序，建议在真机上运行）。

图 10-9 使用 VideoView 组件播放视频文件的案例应用程序的运行效果

下面构建该案例应用程序。在 Android Studio 中新建一个名为 Ch1003 的 Android 应用程序工程。修改 res/layout/activity_main.xml 文件中的代码，使其显示一个 VideoView 组件，修改后的代码如下：

```xml
<RelativeLayout xmlns:android="http://schemas.android.com/apk/res/android"
    android:layout_width="match_parent"
    android:layout_height="match_parent">

    <VideoView
        android:id="@+id/id_videoview"
        android:layout_width="match_parent"
        android:layout_height="match_parent"
        android:layout_centerInParent="true" />

</RelativeLayout>
```

修改 MainActivity.java 文件中的代码，修改后的代码如下：

```java
package com.example.ch1003;

import androidx.appcompat.app.AppCompatActivity;

import android.os.Bundle;
import android.os.Environment;
import android.widget.VideoView;

public class MainActivity extends AppCompatActivity {

    @Override
    protected void onCreate(Bundle savedInstanceState) {
        super.onCreate(savedInstanceState);
        setContentView(R.layout.activity_main);
```

```
        VideoView vv = (VideoView)this.findViewById(R.id.id_videoview);
        String file = Environment.getExternalStorageDirectory()
                                    .getAbsolutePath() + "/bob.mp4";
        System.out.println(file);
        vv.setVideoPath(file);
        vv.start();
    }
}
```

在 MainActivity 类的 onCreate()回调方法中，首先从界面中获取 VideoView 组件的引用，然后设置该 VideoView 组件要播放的视频文件，最后播放该视频文件。

此外，需要在 AndroidManifest.xml 文件中为应用程序访问 SD 卡存储器授权，代码如下：

```
<uses-permission android:name="android.permission.READ_EXTERNAL_STORAGE"/>
```

10.4 同步练习二

修改 10.2 节中的应用程序，使其既可以播放音频文件，又可以播放视频文件。

提示

在判断文件为视频文件后，可以在一个新的 Activity 中使用 VideoView 组件播放视频文件。

第 11 章

存储程序数据

在应用程序的运行过程中，经常需要存储一些程序运行过程中的数据（简称程序数据）。例如，为了避免用户在每次登录时都被要求输入用户名和密码，在用户运行应用程序，首次成功登录后，应用程序会将用户的登录数据保存下来，当用户再次运行该应用程序时，就无须再次输入用户名、密码了。在一般情况下，应用程序都允许用户设置一些程序参数，如网络数据刷新时间、程序字体、每次显示的数据总量等。一旦用户设置了这些程序参数，应用程序就会将这些设置保存下来，当用户再次运行该应用程序时，即可自动采用这些已经设置好的程序参数。

我们可以使用 Android 提供的 SharedPreferences 存储程序数据，也可以使用普通文件存储程序数据，当然，我们还可以将程序数据存储于 SQLite 数据库中。

11.1 使用 SharedPreferences 存储程序数据

SharedPreferences 提供了一个基本框架，通过使用 SharedPreferences，可以非常方便地存储程序数据。想要得到一个 SharedPreferences 对象，只需要在 Activity 中调用 getSharedPreferences(String name, int mode)方法或 getPreferences(int mode)方法，其中的 name 参数是指定的用于存储程序数据的文件名，将 mode 参数的值设置为 0 即可。getPreferences(int mode)方法只是简单地封装了 getSharedPreferences(String name, int mode)方法，并且设置 name 参数为固定的值，即调用这个方法的 Activity 的类名。

下面举例说明如何使用 SharedPreferences 存储程序数据，该案例应用程序运行效果的首界面如图 11-1 所示。首先显示一个登录界面，提示用户输入用户名和密码，如果用户点击"登录"按钮，那么应用程序会将用户输入的用户名和密码保存下来，当用户再次运行该应用程序时，该应用程序会自动地将上次用户输入的用户名和密码填入相应的输入框。

现在构建该案例应用程序。在 Android Studio 中新建一个名为 Ch1101 的 Android 应用程序工程。修改布局文件 res/layout/activity_main.xml 中的代码，修改后

图 11-1 使用 SharedPreferences 存储程序数据的案例应用程序运行效果的首界面

的代码如下：

```xml
<LinearLayout xmlns:android="http://schemas.android.com/apk/res/android"
    android:layout_width="match_parent"
    android:layout_height="match_parent"
    android:orientation="vertical">

    <EditText
        android:id="@+id/id_et_name"
        android:layout_width="match_parent"
        android:layout_height="wrap_content"
        android:hint="@string/text_name"
        />

    <EditText
        android:id="@+id/id_et_password"
        android:layout_width="match_parent"
        android:layout_height="wrap_content"
        android:inputType="textPassword"
        android:hint="@string/text_password"
        />

    <LinearLayout
        android:layout_width="match_parent"
        android:layout_height="wrap_content"
        android:orientation="horizontal"
        >

        <Button
            android:id="@+id/id_btn_login"
            android:layout_width="0dp"
            android:layout_height="wrap_content"
            android:layout_weight="1"
            android:text="@string/text_login"
            />

        <Button
            android:id="@+id/id_btn_reset"
            android:layout_width="0dp"
            android:layout_height="wrap_content"
            android:layout_weight="1"
            android:text="@string/text_reset"
```

```
            />

        </LinearLayout>

</LinearLayout>
```

这个布局文件很简单,该文件中包含两个 EditText 组件和两个 Button 组件。

修改 res/values/strings.xml 文件中的代码,在其中定义布局文件中引用的字符串资源,修改后的代码如下:

```xml
<resources>
    <string name="app_name">Ch1101</string>

    <string name="text_name">用户名</string>
    <string name="text_password">密码</string>
    <string name="text_login">登录</string>
    <string name="text_reset">重置</string>

</resources>
```

修改 MainActivity.java 文件中的代码,使其显示主界面,并且监听对按钮的点击事件,如果用户点击"登录"按钮,则将用户输入的用户名和密码通过 SharedPreferences 保存下来。修改后的 MainActivity.java 文件中的代码如下:

```java
package com.example.ch1101;

import androidx.appcompat.app.AppCompatActivity;

import android.content.SharedPreferences;
import android.os.Bundle;
import android.view.View;
import android.widget.Button;
import android.widget.EditText;

public class MainActivity extends AppCompatActivity implements View.OnClickListener{
    private Button btn_login, btn_reset;
    private EditText et_name, et_password;

    @Override
    protected void onCreate(Bundle savedInstanceState) {
        super.onCreate(savedInstanceState);
        setContentView(R.layout.activity_main);

        btn_login = (Button)this.findViewById(R.id.id_btn_login);
```

```
        btn_login.setOnClickListener(this);
        btn_reset = (Button)this.findViewById(R.id.id_btn_reset);
        btn_reset.setOnClickListener(this);

        et_name = (EditText)this.findViewById(R.id.id_et_name);
        et_password = (EditText)this.findViewById(R.id.id_et_password);

        SharedPreferences sp = this.getSharedPreferences("mimi",0);
        String name = sp.getString("name", "");
        et_name.setText(name);
        String password = sp.getString("password", "");
        et_password.setText(password);
    }

    @Override
    public void onClick(View v) {
        int id = v.getId();
        if (id == R.id.id_btn_login) {
            SharedPreferences sp = this.getSharedPreferences("mimi",0);
            SharedPreferences.Editor editor = sp.edit();
            editor.putString("name", et_name.getText().toString());
            editor.putString("password", et_password.getText().toString());
            editor.commit();
        }
        else {
            et_name.setText("");
            et_password.setText("");
        }
    }
}
```

在 MainActivity 类的 onCreate()回调方法中，首先显示主界面，然后获取界面中两个 EditText 组件及两个 Button 组件的引用，再设置监听按钮点击事件的接口，接下来通过以下代码获取一个 SharedPreferences 对象 sp。

```
SharedPreferences sp = this.getSharedPreferences("mimi",0);
```

注意 getSharedPreferences("mimi", 0)方法的功能：它会在应用程序所在安装目录的特定子目录（/data/data/com.example.ch1101/ shared_prefs 目录）下，检查是否存在 mimi.xml 文件，如果存在，则基于该文件中已有的内容创建一个 SharedPreferences 对象。

最后，通过以下代码从 sp 对象中获取指定 key 对应的 value，即分别获取 key 为"name"和 key 为"password"的 value。如果第一次运行该应用程序，则不能获取相应的 value，因此，getString("name", "")方法和 getString("password","")方法都会返回第二个参数，即空字符串。

```
        String name = sp.getString("name", "");
        et_name.setText(name);
        String password = sp.getString("password", "");
        et_password.setText(password);
```

下面来看对按钮点击事件的响应方法 onClick()，当用户点击"登录"按钮时，通过以下代码获取一个 SharedPreferences 对象 sp，然后调用 sp 对象的 edit()方法，也就是告诉 sp 对象，应用程序要向 sp 对象写入 key/value 对，进而通过 Editor 对象 editor 调用 putString()方法，将用户在输入框中输入的用户名和密码以 key 为"name"和"password"的 value 写入 sp 对象。

```
        SharedPreferences sp = this.getSharedPreferences("mimi",0);
        Editor editor = sp.edit();
        editor.putString("name", et_name.getText().toString());
        editor.putString("password", et_password.getText().toString());
        editor.commit();
```

运行 Ch1101 应用程序，在首界面的两个输入框中分别输入用户名和密码，点击"登录"按钮，然后点击 Android 的退出键退出该应用程序。再次运行该应用程序，即可在首界面的输入框中分别显示上次用户输入的用户名和密码。

在本质上，Android 的 SharedPreferences 就是在应用程序安装目录下的 shared_prefs 子目录下创建的以 getSharedPreferences()方法的第一个参数为文件名的特定的 XML 文件。在 Android Studio 的 Device File Explorer 面板中可以清晰地看到应用程序创建的 XML 文件。在该应用程序中，SharedPreferences 对应的 XML 文件如图 11-2 所示。

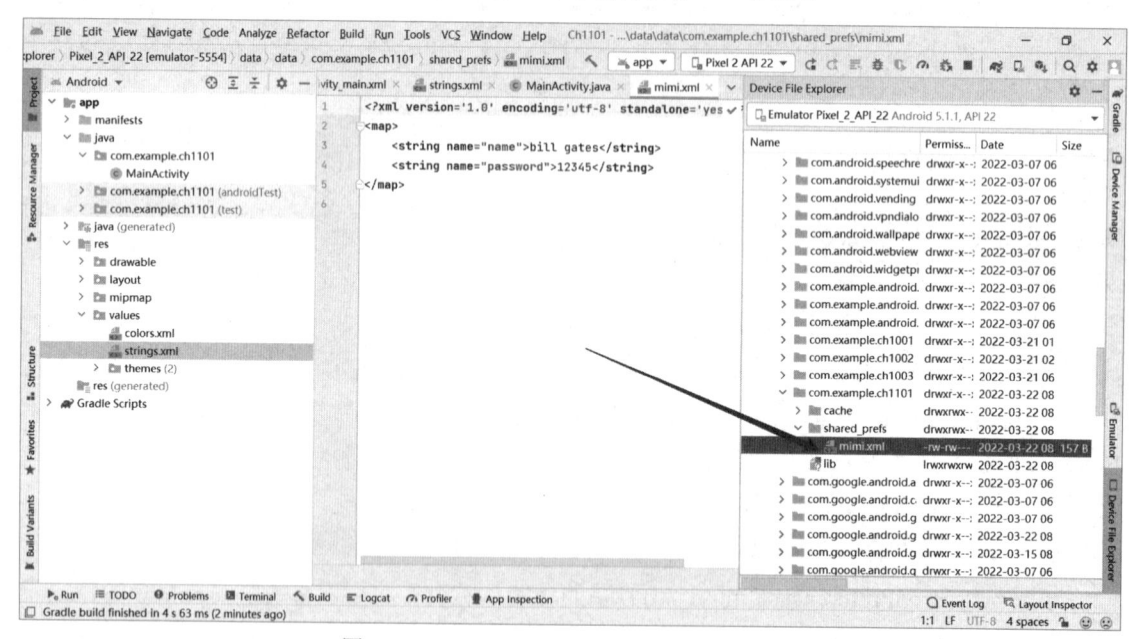

图 11-2 SharedPreferences 对应的 XML 文件

双击 mimi.xml 选项，即可打开 mimi.xml 文件，该文件中的代码如下：

```
<?xml version='1.0' encoding='utf-8' standalone='yes' ?>
<map>
<string name="password">12345</string>
```

```
<string name="name">bill gates</string>
</map>
```

其中的"bill gates"和"12345"分别是在首界面的输入框中输入的用户名和密码。

11.2 同步练习一

完善 11.1 节中的 Ch1101 应用程序，并且对其进行以下修改：当用户点击"登录"按钮时，使用 SharedPreferences 存储用户输入的用户名和密码，以及用户登录的日期和时间。当用户再次运行该应用程序时，不仅会将上次用户输入的用户名和密码自动填入相应的输入框，还会通过一个 Toast 组件显示上次登录的日期和时间。在修改完成后，运行该应用程序，在 Device File Explorer 面板中观察应用程序安装目录下的文件结构变化。

11.3 设置应用程序的首选项

Android 应用程序通常会包括一些设置，以便用户可以改变应用程序的运行特征，如应用程序的界面风格、网络的刷新时间间隔或其他与应用程序有关的运行参数。为了使 Android 应用程序与 Android 自带的系统设置应用程序具有一致的外观，Android 提供了一个名为 Preference 的 API，可以使用 Preference API 设置应用程序的首选项。

一个典型的应用程序首选项设置界面如图 11-3 所示，该界面中包含各种可能的设置。例如，复选框表示是否勾选某个特征，具体值表示可以设置某个参数的值，等等。

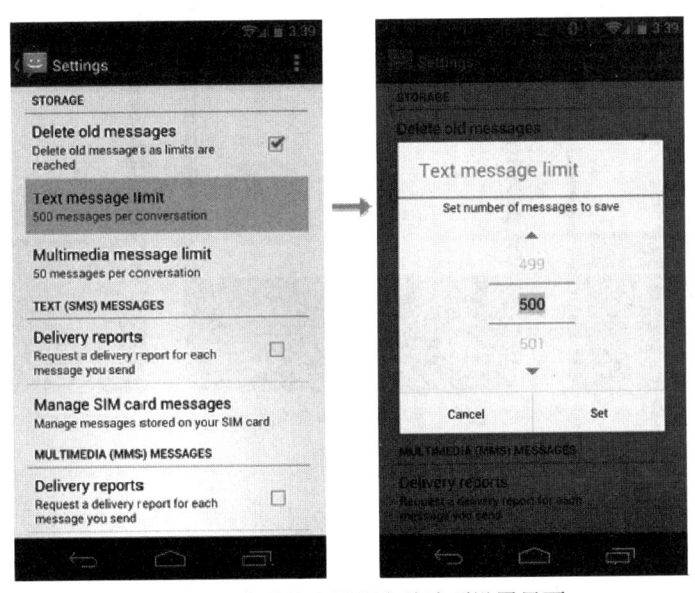

图 11-3　典型的应用程序首选项设置界面

Preference API 提供了应用程序首选项设置界面中所需的所有组件，这些首选项组件所设置的值都以 key/value 对的方式存储于应用程序安装目录下的 shared_prefs 子目录下文件名为"程序包名_preferencex.xml"的 XML 文件中，首选项中存储的值的数据类型包括 Boolean、Float、Int、Long、String 及 String 数组。常用的 Preference 组件如下。

- CheckBoxPreference：显示一个复选框，用于设置某个特征是否可用。
- ListPreference：显示一个列表框，并且在列表框中显示一组单选按钮，用于选择某个特征。
- EditTextPreference：显示一个输入框，用于输入某个值。

下面举例说明如何设置和使用应用程序首选项：使用应用程序首选项设置文字"Hello world!"的样式，可设置的首选项包括是否以动画形式显示界面文本、文本颜色、文本字号、文本的语种（中文、英文），运行效果的首界面如图11-4所示。

点击界面下方的SETTINGS按钮，会显示一个用于设置应用程序首选项的界面，即应用程序首选项设置界面，如图11-5所示。

在应用程序首选项设置界面中可以设置应用程序的各个首选项。例如，如果将文本颜色设置为黑色，然后点击返回按钮回到主界面，那么应用程序主界面中的文本颜色会变为黑色，如图11-6所示。

图 11-4　案例应用程序运行效果的首界面　　图 11-5　应用程序首选项设置界面　　图 11-6　文本颜色变为黑色

现在构建该案例应用程序。在 Android Studio 中新建一个名为 Ch1102 的 Android 应用程序工程。修改布局文件 res/layout/activity_main.xml 中的代码，使其中包含一个 TextView 组件和一个 Button 组件，修改后的代码如下：

```
<LinearLayout xmlns:android="http://schemas.android.com/apk/res/android"
    android:layout_width="match_parent"
    android:layout_height="match_parent"
    android:orientation="vertical">
```

```xml
<TextView
    android:id="@+id/id_textview"
    android:layout_width="match_parent"
    android:layout_height="0dp"
    android:layout_weight="9"
    android:gravity="center"
    android:text="@string/text_hello_world_en" />

<Button
    android:id="@+id/id_btn"
    android:layout_width="match_parent"
    android:layout_height="0dp"
    android:layout_weight="1"
    android:text="@string/setting" />
```
```
</LinearLayout>
```

为了能够显示应用程序首选项设置界面，需要在 res 目录下新建一个 xml 子目录，并且在 res/xml 目录下新建一个名为 my_settings.xml 的首选项文件（文件名可以是任意合法的文件名），修改该文件中的代码，修改后的代码如下：

```xml
<?xml version="1.0" encoding="utf-8"?>
<PreferenceScreen xmlns:android="http://schemas.android.com/apk/res/android">
    <CheckBoxPreference
        android:key="pref_animation"
        android:title="@string/text_pref_animation"
        android:summary="@string/text_pref_animation_summ"
        android:defaultValue="false" />
    <ListPreference
        android:key="pref_lang_type"
        android:title="@string/text_pref_lang_type"
        android:dialogTitle="@string/text_pref_lang_type"
        android:entries="@array/pref_lang_type_entries"
        android:entryValues="@array/pref_lang_type_values"
        android:defaultValue="1" />
    <PreferenceCategory
        android:title="@string/text_pref_appearence_title">
        <EditTextPreference
            android:key="pref_text_size"
            android:title="@string/text_pref_text_size"
            android:summary="@string/text_pref_text_size_summ"
            android:defaultValue="16"/>
        <EditTextPreference
```

```
            android:key="pref_text_color"
            android:title="@string/text_pref_text_color"
            android:summary="@string/text_pref_text_color_summ"
            android:defaultValue="#FF000000"/>
    </PreferenceCategory>

</PreferenceScreen>
```

在 my_settings.xml 文件中，使用<PreferenceScreen>标签表示要构建一个应用程序首选项设置界面，然后在该标签下，通过以下代码构建一个设置是否显示动画的选项。

```
    <CheckBoxPreference
        android:key="pref_animation"
        android:title="@string/text_pref_animation"
        android:summary="@string/text_pref_animation_summ"
        android:defaultValue="false" />
```

其中，以下两条代码主要用于设置该选项的 key 和默认值，即 key/value 对。

```
        android:key="pref_animation"
```

和

```
        android:defaultValue="false" />
```

此外，通过 title 属性和 summary 属性设置在应用程序首选项设置界面中显示的提示信息。类似地，通过以下代码设置一个用于选择文字显示语种的列表选项。

```
    <ListPreference
        android:key="pref_lang_type"
        android:title="@string/text_pref_lang_type"
        android:dialogTitle="@string/text_pref_lang_type"
        android:entries="@array/pref_lang_type_entries"
        android:entryValues="@array/pref_lang_type_values"
        android:defaultValue="1" />
```

其中，以下两条代码主要用于设置 key/value 对。

```
        android:key="pref_lang_type"
```

和

```
        android:defaultValue="1" />
```

以下代码分别用于设置在应用程序首选项设置界面中显示的提示信息、打开列表框时显示的提示信息、列表框中显示的列表信息及其对应的选项值。

```
            android:title="@string/text_pref_lang_type"
            android:dialogTitle="@string/text_pref_lang_type"
            android:entries="@array/pref_lang_type_entries"
            android:entryValues="@array/pref_lang_type_values"
```

由于文本字号和文本颜色都是文本外观，因此，需要对这两项设置进行分组：会将它们显示在一个分组中。以下代码主要用于设置分组和分组显示的名称。

```xml
<PreferenceCategory
    android:title="@string/text_pref_appearence_title">
```

在这个分组中设置两个输入框,分别用于设置文本字号和文本颜色,其方法与设置动画属性的方法类似。

在 activity_main.xml 文件和 my_settings.xml 文件中引用了一些字符串资源和字符串数组资源,因此,需要修改 res/values/strings.xml 文件中的代码,在其中定义需要引用的字符串资源和字符串数组资源,修改后的代码如下:

```xml
<resources>
    <string name="app_name">Ch1102</string>

    <string name="setting">Settings</string>

    <string name="text_hello_world_en">Hello world!</string>
    <string name="text_hello_world_cn">你好,世界!</string>

    <string name="text_pref_animation">以动画形式显示文字</string>
    <string name="text_pref_animation_summ">选中该选项,
                        可以以动画的形式显示界面文字</string>
    <string name="text_pref_lang_type">选择显示文字的语种</string>

    <string-array name="pref_lang_type_entries">
        <item>English</item>
        <item>中文</item>
    </string-array>

    <string-array name="pref_lang_type_values">
        <item>1</item>
        <item>2</item>
    </string-array>

    <string name="text_pref_appearence_title">设置文字显示的外观</string>
    <string name="text_pref_text_size">文本字号</string>
    <string name="text_pref_text_size_summ">文本字号,请输入一个数值</string>
    <string name="text_pref_text_color">文本颜色</string>
 <string name="text_pref_text_color_summ">文本颜色,请输入一个数值,
                        格式:#AARRGGBB</string>

</resources>
```

为了显示动画,我们需要在 res 目录下新建一个 anim 子目录,并且在 res/anim 目录下新建一个名为 my_scale.xml 的动画定义文件,修改该文件中的代码,修改后的代码如下:

```xml
<?xml version="1.0" encoding="utf-8"?>
<scale xmlns:android="http://schemas.android.com/apk/res/android"
```

```xml
    android:interpolator="@android:anim/accelerate_interpolator"
    android:fromXScale="0"
    android:toXScale="1"
    android:fromYScale="0"
    android:toYScale="1"
    android:duration="4000"
    android:pivotX="50%"
    android:pivotY="50%"
    android:fillAfter="true">
</scale>
```

现在构建应用程序首选项设置界面的 Activity，在 src 目录下的 com.example.ch1102 包中新建一个名为 MySettingsActivity 的 Java 类文件。修改 MySettingsActivity.java 文件中的代码，修改后的代码如下：

```java
package com.example.ch1102;

import android.os.Bundle;
import android.preference.PreferenceActivity;

public class MySettingsActivity extends PreferenceActivity {
    public static String PREF_ANIMATION = "pref_animation";
    public static String PREF_LANG_TYPE = "pref_lang_type";
    public static String PREF_TEXT_SIZE = "pref_text_size";
    public static String PREF_TEXT_COLOR = "pref_text_color";

    @SuppressWarnings("deprecation")
    @Override
    public void onCreate(Bundle savedInstanceState) {
        super.onCreate(savedInstanceState);
        addPreferencesFromResource(R.xml.my_settings);
    }
}
```

在 MySettingsActivity.java 文件中定义了几个与 my_settings.xml 文件中同名的字符串资源，用于设置列表选项的 key，代码如下：

```java
    public static String PREF_ANIMATION = "pref_animation";
    public static String PREF_LANG_TYPE = "pref_lang_type";
    public static String PREF_TEXT_SIZE = "pref_text_size";
    public static String PREF_TEXT_COLOR = "pref_text_color";
```

通过这几个 key 可以获取在应用程序首选项设置界面中设置的值。

下面看一下 MySettingsActivity 类中的 onCreate() 回调方法，代码如下：

```java
    public void onCreate(Bundle savedInstanceState) {
        super.onCreate(savedInstanceState);
```

```
        addPreferencesFromResource(R.xml.my_settings);
    }
```
通过以下方法可以显示应用程序首选项设置界面，并且用户可以修改应用程序首选项的值。
```
        addPreferencesFromResource(R.xml.my_settings);
```
现在返回 MainActivity 类，看一下如何获取和使用在 MySettingsActivity 类中设置的应用程序首选项值。修改 MainActivity.java 文件中的代码，修改后的代码如下：

```
package com.example.ch1102;

import androidx.appcompat.app.AppCompatActivity;

import android.content.Intent;
import android.content.SharedPreferences;
import android.graphics.Color;
import android.os.Bundle;
import android.preference.PreferenceManager;
import android.view.View;
import android.view.animation.Animation;
import android.view.animation.AnimationUtils;
import android.widget.Button;
import android.widget.TextView;

public class MainActivity extends AppCompatActivity implements
        SharedPreferences.OnSharedPreferenceChangeListener, View.OnClickListener {
    private TextView tv;

    @Override
    protected void onCreate(Bundle savedInstanceState) {
        super.onCreate(savedInstanceState);
        setContentView(R.layout.activity_main);

        Button btn = findViewById(R.id.id_btn);
        btn.setOnClickListener(this);

        tv = (TextView) findViewById(R.id.id_textview);

        SharedPreferences settings =
                PreferenceManager.getDefaultSharedPreferences(this);
        settings.registerOnSharedPreferenceChangeListener(this);

        String lang_type =
                settings.getString(MySettingsActivity.PREF_LANG_TYPE, "1");
        if (lang_type.equalsIgnoreCase("1"))
```

```java
            tv.setText(R.string.text_hello_world_en);
        else
            tv.setText(R.string.text_hello_world_cn);

        String text_size =
                settings.getString(MySettingsActivity.PREF_TEXT_SIZE, "16");
        tv.setTextSize(Float.valueOf(text_size));

        String text_color = settings.getString(MySettingsActivity.PREF_TEXT_COLOR,
                "#FF000000");
        tv.setTextColor(Color.parseColor(text_color));

        Boolean animation = settings.getBoolean(MySettingsActivity.PREF_ANIMATION,
                false);
        if (animation == true) {
            Animation scale = AnimationUtils.loadAnimation(this, R.anim.my_scale);
            tv.startAnimation(scale);
        }
    }

    @Override
    public void onSharedPreferenceChanged(SharedPreferences settings, String key) {
        if (key.equalsIgnoreCase(MySettingsActivity.PREF_ANIMATION)) {
            Boolean animation = settings.getBoolean(
                    MySettingsActivity.PREF_ANIMATION, false);
            if (animation == true) {
                Animation scale = AnimationUtils.loadAnimation(this, R.anim.my_scale);
                tv.startAnimation(scale);
            }
        }
        else if (key.equalsIgnoreCase(MySettingsActivity.PREF_LANG_TYPE)) {
            String lang_type = settings.getString(
                    MySettingsActivity.PREF_LANG_TYPE, "1");
            if (lang_type.equalsIgnoreCase("1"))
                tv.setText(R.string.text_hello_world_en);
            else
                tv.setText(R.string.text_hello_world_cn);
        }
        else if (key.equalsIgnoreCase(MySettingsActivity.PREF_TEXT_COLOR)) {
            String text_color = settings.getString(
                    MySettingsActivity.PREF_TEXT_COLOR, "#FFFFFFFF");
            tv.setTextColor(Color.parseColor(text_color));
        }
```

```
        else if (key.equalsIgnoreCase(MySettingsActivity.PREF_TEXT_SIZE)) {
            String text_size =
                    settings.getString(MySettingsActivity.PREF_TEXT_SIZE, "16");
            tv.setTextSize(Float.valueOf(text_size));
        }
    }



    @Override
    public void onClick(View view) {
        Intent i = new Intent(this, MySettingsActivity.class);
        this.startActivity(i);
    }
}
```

在 MainActivity 类的 onCreate()回调方法中，首先显示应用程序界面并获取 TextView 组件的引用；然后通过以下代码获取应用程序首选项对象，并且设置应用程序首选项中的参数值被改变时的监听方法：无论应用程序首选项中的哪个参数值被改变，都会调用该接口的 onSharedPreferenceChanged()方法，用于改变文本框的显示特性。

```
        SharedPreferences settings =
                    PreferenceManager.getDefaultSharedPreferences(this);
        settings.registerOnSharedPreferenceChangeListener(this);
```

最后，通过以下代码从应用程序首选项对象中获取各个选项设置的参数值，并且根据所选择的参数改变文本框的显示特性。

```
        String lang_type =
                settings.getString(MySettingsActivity.PREF_LANG_TYPE, "1");
        if (lang_type.equalsIgnoreCase("1"))
            tv.setText(R.string.text_hello_world_en);
        else
            tv.setText(R.string.text_hello_world_cn);

        String text_size =
                settings.getString(MySettingsActivity.PREF_TEXT_SIZE, "16");
        tv.setTextSize(Float.valueOf(text_size));

        String text_color = settings.getString(MySettingsActivity.PREF_TEXT_COLOR,
                                               "#FF000000");
        tv.setTextColor(Color.parseColor(text_color));

        Boolean animation = settings.getBoolean(MySettingsActivity.PREF_ANIMATION,
                                                false);
        if (animation == true) {
```

```
        Animation scale = AnimationUtils.loadAnimation(this, R.anim.my_scale);
        tv.startAnimation(scale);
    }
```

现在看一下 onSharedPreferenceChanged()方法，当应用程序首选项中的任意一个参数值被改变时，都会调用该方法。onSharedPreferenceChanged()方法会根据被改变的参数值实时改变文本框的显示特性，代码如下：

```
@Override
public void onSharedPreferenceChanged(SharedPreferences settings, String key) {
    if (key.equalsIgnoreCase(MySettingsActivity.PREF_ANIMATION)) {
        Boolean animation = settings.getBoolean(
                        MySettingsActivity.PREF_ANIMATION, false);
        if (animation == true) {
            Animation scale =
                    AnimationUtils.loadAnimation(this, R.anim.my_scale);
            tv.startAnimation(scale);
        }
    }
    else if (key.equalsIgnoreCase(MySettingsActivity.PREF_LANG_TYPE)) {
        String lang_type = settings.getString(
                        MySettingsActivity.PREF_LANG_TYPE, "1");
        if (lang_type.equalsIgnoreCase("1"))
            tv.setText(R.string.text_hello_world_en);
        else
            tv.setText(R.string.text_hello_world_cn);
    }
    else if (key.equalsIgnoreCase(MySettingsActivity.PREF_TEXT_COLOR)) {
        String text_color = settings.getString(
                        MySettingsActivity.PREF_TEXT_COLOR, "#FFFFFFFF");
        tv.setTextColor(Color.parseColor(text_color));
    }
    else if (key.equalsIgnoreCase(MySettingsActivity.PREF_TEXT_SIZE)) {
        String text_size =
                settings.getString(MySettingsActivity.PREF_TEXT_SIZE, "16");
        tv.setTextSize(Float.valueOf(text_size));
    }
}
```

此外，需要在 AndroidManifest.xml 文件中配置 MySettingsActivity 类，代码如下：

```
<?xml version="1.0" encoding="utf-8"?>
<manifest xmlns:android="http://schemas.android.com/apk/res/android"
    package="com.example.ch1102">

    <application
```

```xml
        android:allowBackup="true"
        android:icon="@mipmap/ic_launcher"
        android:label="@string/app_name"
        android:roundIcon="@mipmap/ic_launcher_round"
        android:supportsRtl="true"
        android:theme="@style/Theme.Ch1102">
        <activity
            android:name=".MainActivity"
            android:exported="true">
            <intent-filter>
                <action android:name="android.intent.action.MAIN" />

                <category android:name="android.intent.category.LAUNCHER" />
            </intent-filter>
        </activity>

        <activity
            android:name=".MySettingsActivity"
            android:exported="false"/>

    </application>

</manifest>
```

运行 Ch1102 应用程序，即可打开图 11-4 所示的界面，点击 SETTINGS 按钮，即可打开图 11-5 所示的界面。

11.4 同步练习二

将 11.3 节中的案例应用程序复制到自己的开发环境中并运行，观察该应用程序的运行效果，并且进行以下修改：在应用程序首选项设置界面中添加一个新的首选项，用于设置动画的执行时间。

提示

在获取动画对象后，使用 setDuration() 方法设置动画的执行时间。由于该首选项只有在执行动画时才有效，因此，需要使用 android:dependency 首选项属性配置这个首选项的依赖。

11.5 在应用程序目录下存储程序数据

任意一个 Android 应用程序，在被安装到 Android 平台上时，Android 平台都会在/data/data 目录下，以该应用程序的包名为名，为该应用程序创建一个唯一的子目录，我们可以在该子目录下创建只有该应用程序才可以访问的子目录或文件。本节主要介绍如何在应用程序的私有目录下创建和使用文件。

Activity 提供了以下用于操作应用程序私有目录的方法。这些方法是在 Context 类中定义的，因为 Activity 类是 Context 类的子类，所以继承了这些方法。

- File Activity.getFilesDir()。返回应用程序私有目录完整路径的 File 对象。例如，如果应用程序的包名为 com.ttt.mysample，那么这个 File 对象的完整路径为/data/data/com.ttt.mysample/files。
- File getDir(String name, int mode)。在应用程序的私有目录下，创建或返回一个名字为"app_" + name 参数的子目录 File 对象，设置 mode 参数的值为 0 即可。
- boolean deleteFile(String name)。在应用程序的私有目录下，删除一个名字与 name 参数一致的文件。
- String[] fileList()。返回应用程序私有目录下的所有子目录和文件名。
- FileOutputStream openFileOutput(String name, int mode)。在应用程序的私有目录下，打开或新建一个名字与 name 参数一致的文件，用于写入数据，设置 mode 参数的值为 0 即可。
- FileInputStream openFileInput(String name)。在应用程序的私有目录下，打开名字与 name 参数一致的文件，用于读取数据。

除了以上方法，还可以使用 java.io 包中提供的类与方法，对应用程序私有目录下的任意文件或目录进行操作。

11.6 同步练习三

参照 11.5 节中介绍的各个文件访问方法的功能，编写一个简单的 Android 应用程序。

提示

简单地输出相关信息即可。

11.7 访问外部存储器

对于配备了外部 SD 卡的 Android 设备，可以使用 Android 提供的 Environment 工具类检查外部 SD 卡的状态及获取 SD 卡中的特定子目录，如是否插入 SD 卡、SD 卡当前是否可读、SD 卡当前是否可写、获取 SD 卡中特定子目录的 File 对象等。

在读/写外部 SD 卡前，需要为应用程序在 AndroidManifest.xml 文件中申请相应的权限。例如，为了读/写 SD 卡，需要编写以下代码。

```
<uses-permission android:name="android.permission.WRITE_EXTERNAL_STORAGE" />
```

如果只需要读取 SD 卡中的数据，则需要编写以下代码。

```
<uses-permission android:name="android.permission.READ_EXTERNAL_STORAGE" />
```

11.7.1 检查 SD 卡的状态

在对外部 SD 卡进行读/写操作前，需要检查 SD 卡的状态，使用 Environment 类中的 getExternalStorageState()方法获取 SD 卡的状态，其状态包括 MEDIA_UNKNOWN、MEDIA_REMOVED、MEDIA_UNMOUNTED、MEDIA_CHECKING、MEDIA_NOFS、MEDIA_MOUNTED、MEDIA_MOUNTED_READ_ONLY、MEDIA_SHARED、MEDIA_BAD_REMOVAL、MEDIA_UNMOUNTABLE，它们的含义是不言自明的，如以下代码片段主要用于检查外部 SD 卡当前的状态是否可读/写。

```java
/* 检查外部SD卡当前的状态是否可写，若是，则返回true，否则返回false */
public boolean isExternalStorageWritable() {
    String state = Environment.getExternalStorageState();
    if (Environment.MEDIA_MOUNTED.equals(state)) {
        return true;
    }
    return false;
}

/* 检查外部SD卡当前的状态是否可读，若是，则返回true，否则返回false */
public boolean isExternalStorageReadable() {
    String state = Environment.getExternalStorageState();
    if (Environment.MEDIA_MOUNTED.equals(state) ||
        Environment.MEDIA_MOUNTED_READ_ONLY.equals(state)) {
        return true;
    }
    return false;
}
```

11.7.2 获取 SD 卡中特定子目录的 File 对象

使用 Environment 类中的 getExternalStoragePublicDirectory(String type)方法可以获取 SD 卡中特定子目录的 File 对象，其中，type 参数的值包括 DIRECTORY_MUSIC、DIRECTORY_PODCASTS、DIRECTORY_RINGTONES、DIRECTORY_ALARMS、DIRECTORY_NOTIFICATIONS、DIRECTORY_PICTURES、DIRECTORY_MOVIES、DIRECTORY_DOWNLOADS、DIRECTORY_DCIM，它们的含义是不言自明的，如以下代码片段主要用于返回 SD 卡中存储的图片资源文件的完整路径和目录名。

```java
String fullpath = Environment.getExternalStoragePublicDirectory(
                Environment.DIRECTORY_PICTURES).getAbsolutePath();
String path = Environment.getExternalStoragePublicDirectory(
                Environment.DIRECTORY_PICTURES).getName();
```

在获取所需子目录后，就可以使用 java.io 包中的相关类及其方法，像操作普通文件一样，操作 SD 卡中的文件或目录了。

11.8 使用 SQLite 数据库存储程序数据

11.8.1 SQLite 数据库简介

SQLite 是一个开源、免费的数据库管理系统。对于基于 C/S 模式的数据库管理系统,如 Microsoft SQL Server、MySQL,数据库存储于称为服务器的计算机系统中,并且通过数据库管理系统的服务器端程序进行管理。需要使用数据库中数据的程序,称为客户端程序。客户端程序通过某种通信协议(如 TCP/IP)与数据库管理系统的服务器端程序通信,从而对数据库中的数据进行操作。SQLite 与基于 C/S 模式的数据库管理系统不同,它不是基于 C/S 模式的,它只是一个 C 语言程序包(C 语言方法库),需要使用 SQLite 数据库的程序通过调用这个程序包中的方法,即可进行创建数据库、访问数据库中数据等操作。

Android 整合了 SQLite 数据库管理系统,对 SQLite 的 C 语言程序包进行了 Java 封装,提供了基于 Java 语言的类库。因此,在 Android 应用程序开发中,可以使用 SQLite 提供的 Java 接口创建及访问 SQLite 数据库。

我们可以使用 SQL 规范中定义的 SQL 语言创建和操作 SQLite 数据库。需要注意的是,对于所支持的数据类型,基于 C/S 模式的数据库管理系统都提供了十分丰富的数据类型,如 int、char、varchar、text、image、real。SQLite 也支持这些数据类型,但是,它会采用 Type Affinity 机制对数据类型进行自动映射,也就是将所有的数据类型都映射到以下 5 种类型中。

- TEXT。字符串类型。
- NUMBERIC。精确表示的数值类型。
- INTEGER。整数类型。
- REAL。采用 8 字节表示的 IEEE 浮点数据类型,与 NUMBERIC 不同,该数据类型可能会有数据精度损失。
- BLOB。二进制数据类型。

读者不需要详细了解 Type Affinity 机制,在创建数据库表时,使用以上 5 种数据类型即可。

11.8.2 在 Android 中使用 SQLite 数据库

Android 提供了操作 SQLite 数据库的 Java 类库,对 SQLite 数据库完全支持,其中一个非常重要和常用的类就是 SQLiteOpenHelper 类,在需要创建和使用 SQLite 数据库时,应该继承该类,并且重写其中的 onCreate()方法,从而创建所需的数据表。

下面举例说明如何使用 SQLiteOpenHelper 类创建 SQLite 数据库,以及如何在数据库中创建所需的数据表。我们要创建的数据库名为 Teach.db,该数据库中包含 3 个数据表,分别为 student 表、course 表和 score 表,分别用于存储学生信息、课程信息和成绩信息,这 3 个数据表的结构定义分别如表 11-1、表 11-2 和表 11-3 所示。

表 11-1 student 表的结构定义

字 段 名 称	数 据 类 型	备 注
_id	INTEGER	学生编号,主键,自增
student_name	TEXT	学生姓名
student_birth	TEXT	学生出生日期
student_phone	TEXT	联系电话
student_photo	BLOB	学生头像

表 11-2　course 表的结构定义

字 段 名 称	数 据 类 型	备　　注
_id	INTEGER	课程编号，主键，自增
course_name	TEXT	课程名称
course_memo	TEXT	课程介绍

表 11-3　score 表的结构定义

字 段 名 称	数 据 类 型	备　　注
_id	INTEGER	成绩编号，主键，自增
student_id	INTEGER	学生编号，外键，参照 student 表中的_id 字段
course_id	INTEGER	课程编号，外键，参照 course 表中的_id 字段
score_score	REAL	学生课程成绩

下面创建一个 Android 应用程序，创建所需的数据库，并且在数据库中创建所需的数据表。在 Android Studio 中新建一个名为 Ch1103 的 Android 应用程序工程，并且在 src 目录下新建一个名为 com.example.ch1103.database 的包，在该包中新建一个名为 MyDataBaseHelper 的 Java 类文件。修改 MyDataBaseHelper.java 文件中的代码，修改后的代码如下：

```java
package com.example.ch1103.database;

import android.content.Context;
import android.database.sqlite.SQLiteDatabase;
import android.database.sqlite.SQLiteOpenHelper;

public class MyDataBaseHelper extends SQLiteOpenHelper {

    private static final String DATABASE_NAME = "Teach.db";
    private static final int DATABASE_VERSION = 1;

    private static final String creat_student = "CREATE TABLE student ( " +
            "_id INTEGER PRIMARY KEY," +
            "student_name TEXT, " +
            "student_birth TEXT, " +
            "student_phone TEXT, " +
            "student_photo BLOB " +
            ");";
    private static final String creat_course = "CREATE TABLE course ( " +
            "_id INTEGER PRIMARY KEY, " +
            "course_name TEXT, " +
            "course_memo TEXT " +
            ");";
    private static final String creat_score = "CREATE TABLE score ( " +
```

```
            "_id INTEGER PRIMARY KEY," +
            "student_id INTEGER , " +
            "course_id INTEGER, " +
            "FOREIGN KEY(student_id) REFERENCES student(student_id), " +
            "FOREIGN KEY(course_id) REFERENCES course(course_id) " +
            ");";


    public MyDataBaseHelper(Context context) {
        super(context, DATABASE_NAME, null, DATABASE_VERSION);
    }

    @Override
    public void onCreate(SQLiteDatabase db) {
        db.execSQL(creat_student);
        db.execSQL(creat_course);
        db.execSQL(creat_score);
    }

    @Override
    public void onUpgrade(SQLiteDatabase db, int oldVersion, int newVersion) {
        db.execSQL("drop table score");
        db.execSQL("drop table student");
        db.execSQL("drop table course");

        onCreate(db);
    }
}
```

在 MyDataBaseHelper 类中定义了数据库的名称、数据库的版本和几个用于创建数据表的字符串常量。首先看一下该类的构造方法，代码如下：

```
    public MyDataBaseHelper(Context context) {
        super(context, DATABASE_NAME, null, DATABASE_VERSION);
    }
```

该构造方法直接调用了父类的构造方法，父类的构造方法会检查是否已经存在名为 Teach.db 且版本号为 1 的数据库，如果不存在，则会自动创建 Teach.db 数据库，并且调用 onCreate()方法创建所需的数据表。在本案例中，onCreate()方法创建了 3 个数据表，分别为 student 表、course 表和 score 表，如果已经存在名为 Teach.db 且版本号也为 1 的数据库，则不进行任何操作；如果已经存在名为 Teach.db 的数据库，但是版本号小于由参数 DATABASE_VERSION 指定的版本号，则会调用 onUpgrade()方法升级数据库。在本案例中，onUpgrade()方法简单地删除了以前的数据表，然后重新创建了所需的数据表。

有了这个数据库操作辅助类，我们就可以轻松地操作数据表中的数据了。

现在继续完善 Ch1103 应用程序：首先向 student 表中插入几条记录，然后使用一个 ListView 组件将 student 表中的数据显示出来。

运行 Ch1103 应用程序，运行效果的首界面如图 11-7 所示。点击"插入并显示学生记录"按钮，首先将 3 条记录插入 student 表，然后从 student 表中将插入的记录读取出来，最后使用 ListView 组件将其显示出来，如图 11-8 所示。

图 11-7　Ch1103 应用程序运行效果的首界面　　图 11-8　插入 3 条记录并在 ListView 组件中显示

student 表中包含学生头像，为了方便，我们直接将表示学生头像的 3 张图片存储于 res/drawable 目录下。现在修改主界面布局文件 res/layout/activity_main.xml 中的代码，修改后的代码如下：

```xml
<LinearLayout xmlns:android="http://schemas.android.com/apk/res/android"
    android:layout_width="match_parent"
    android:layout_height="match_parent"
    android:orientation="vertical">

    <Button
        android:id="@+id/id_button"
        android:layout_width="match_parent"
        android:layout_height="wrap_content"
        android:text="@string/text_button" />

    <ListView
        android:id="@+id/id_listview"
```

```xml
        android:layout_width="match_parent"
        android:layout_height="wrap_content"
        />

</LinearLayout>
```

这个布局文件很简单,主要包含一个 Button 组件和一个 ListView 组件。

在 res/layout 目录下新建一个列表项布局文件 list_item.xml,修改该文件中的代码,在 ListView 组件的列表项中显示学生头像、学生编号、学生姓名、学生出生日期和联系电话,修改后的代码如下:

```xml
<?xml version="1.0" encoding="utf-8"?>
<LinearLayout xmlns:android="http://schemas.android.com/apk/res/android"
    android:layout_width="match_parent"
    android:layout_height="match_parent"
    android:orientation="horizontal" >

    <ImageView
        android:id="@+id/id_li_photo"
        android:layout_width="64dp"
        android:layout_height="64dp"
        android:scaleType="fitCenter"
        android:contentDescription="@string/hello_world"
        />

    <View
        android:layout_width="10dp"
        android:layout_height="64dp"
        />

    <LinearLayout
        android:layout_width="match_parent"
        android:layout_height="wrap_content"
        android:orientation="vertical"
        >

        <TextView
            android:id="@+id/id_li_id"
            android:layout_width="match_parent"
            android:layout_height="wrap_content"
            />

        <TextView
            android:id="@+id/id_li_name"
```

```xml
            android:layout_width="match_parent"
            android:layout_height="wrap_content"
            />

        <TextView
            android:id="@+id/id_li_birth"
            android:layout_width="match_parent"
            android:layout_height="wrap_content"
            />

        <TextView
            android:id="@+id/id_li_phone"
            android:layout_width="match_parent"
            android:layout_height="wrap_content"
            />

    </LinearLayout>

</LinearLayout>
```

修改 res/values/strings.xml 文件中的代码，在其中定义布局文件中引用的字符串资源，修改后的代码如下：

```xml
<resources>
    <string name="app_name">Ch1103</string>

    <string name="text_button">插入并显示学生记录</string>
    <string name="hello_world">Hello world!</string>
</resources>
```

为了能在 ListView 组件中显示学生信息，需要构建我们自己的 Adapter。在 src 目录下新建一个名为 com.example.ch1103.adapters 的包，并且在该包中新建一个名为 MySimpleCursorAdapter 的 Java 类文件。修改 MySimpleCursorAdapter.java 文件中的代码，修改后的代码如下：

```java
package com.example.ch1103.adapters;

import android.content.Context;
import android.database.Cursor;
import android.graphics.Bitmap;
import android.graphics.BitmapFactory;
import android.view.LayoutInflater;
import android.view.View;
import android.view.ViewGroup;
import android.widget.ImageView;
import android.widget.SimpleCursorAdapter;
import android.widget.TextView;
```

```java
public class MySimpleCursorAdapter extends SimpleCursorAdapter {
    private int layout;
    private Cursor c;
    private String[] from;
    private int[] to;

    private LayoutInflater mInflater;

    @SuppressWarnings("deprecation")
    public MySimpleCursorAdapter(Context context, int layout, Cursor c,
                        String[] from, int[] to) {
        super(context, layout, c, from, to);

        this.layout = layout;
        this.c = c;
        this.from = from;
        this.to = to;

        mInflater = (LayoutInflater) context
                .getSystemService(Context.LAYOUT_INFLATER_SERVICE);
    }

    @Override
    public View getView(int position, View convertView, ViewGroup parent) {
        View v;
        v = mInflater.inflate(layout, parent, false);

        ImageView iv_photo = (ImageView) v.findViewById(to[0]);
        int t = c.getColumnIndex(from[0]);
        byte[] photo = c.getBlob(t);
        Bitmap bm = BitmapFactory.decodeByteArray(photo, 0, photo.length);
        iv_photo.setImageBitmap(bm);

        TextView id = (TextView) v.findViewById(to[1]);
        t = c.getColumnIndex(from[1]);
        String d = c.getString(t);
        id.setText(d);

        TextView name = (TextView) v.findViewById(to[2]);
        t = c.getColumnIndex(from[2]);
        d = c.getString(t);
        name.setText(d);
```

```java
            TextView birth = (TextView) v.findViewById(to[3]);
            t = c.getColumnIndex(from[3]);
            d = c.getString(t);
            birth.setText(d);

            TextView phone = (TextView) v.findViewById(to[4]);
            t = c.getColumnIndex(from[4]);
            d = c.getString(t);
            phone.setText(d);

            return v;

        }

}
```

在 MySimpleCursorAdapter 类的 getView()方法中，可以通过 Cursor 对象获取学生头像的二进制数据，并且通过 BitmapFactory 将学生头像的二进制数据组装成 Bitmap 对象，然后将其显示在 ImageView 组件中。对于其他字符串数据，可以直接显示在相应的 TextView 组件中。

修改 MainActivity.java 文件中的代码，使其显示主界面，监听对按钮的点击事件，并且在对按钮点击事件的响应方法中，将 3 条学生信息记录插入 student 表，然后从 student 表中将插入的记录读取出来，并且将其显示在 ListView 组件中，修改后的代码如下：

```java
package com.example.ch1103;

import androidx.appcompat.app.AppCompatActivity;

import android.content.ContentValues;
import android.database.Cursor;
import android.database.sqlite.SQLiteDatabase;
import android.graphics.Bitmap;
import android.graphics.BitmapFactory;
import android.os.Bundle;
import android.view.View;
import android.widget.Button;
import android.widget.ListView;

import com.example.ch1103.adapters.MySimpleCursorAdapter;
import com.example.ch1103.database.MyDataBaseHelper;

import java.io.ByteArrayOutputStream;
import java.io.IOException;
```

```java
public class MainActivity extends AppCompatActivity implements View.OnClickListener {
    private Button btn;
    private ListView lv;

    @Override
    protected void onCreate(Bundle savedInstanceState) {
        super.onCreate(savedInstanceState);
        setContentView(R.layout.activity_main);

        btn = (Button)this.findViewById(R.id.id_button);
        btn.setOnClickListener(this);

        lv = (ListView)this.findViewById(R.id.id_listview);
    }

    @Override
    public void onClick(View v) {
        MyDataBaseHelper helper = new MyDataBaseHelper(this);
        SQLiteDatabase dbw = helper.getWritableDatabase();

        ContentValues values = new ContentValues();
        ByteArrayOutputStream baos;
        Bitmap photo;

        //插入3条学生信息记录
        values.put("student_name", "张大卫");
        values.put("student_birth", "2011-01-01");
        values.put("student_phone", "13800138000");
        photo = BitmapFactory.decodeResource(this.getResources(), R.drawable.png0010);
        baos = new ByteArrayOutputStream();
        photo.compress(Bitmap.CompressFormat.PNG, 100, baos);
        values.put("student_photo", baos.toByteArray());
        try {
            baos.close();
        } catch (IOException e) {
            e.printStackTrace();
        }
        dbw.insert("student", null, values);

        values.put("student_name", "李丹丹");
        values.put("student_birth", "2001-11-09");
        values.put("student_phone", "13800138001");
```

```
        photo = BitmapFactory.decodeResource(this.getResources(), R.drawable.png0015);
        baos = new ByteArrayOutputStream();
        photo.compress(Bitmap.CompressFormat.PNG, 100, baos);
        values.put("student_photo", baos.toByteArray());
        try {
            baos.close();
        } catch (IOException e) {
            e.printStackTrace();
        }
        dbw.insert("student", null, values);

        values.put("student_name", "王芳");
        values.put("student_birth", "1990-10-10");
        values.put("student_phone", "13800138002");
        photo = BitmapFactory.decodeResource(this.getResources(), R.drawable.png1783);
        baos = new ByteArrayOutputStream();
        photo.compress(Bitmap.CompressFormat.PNG, 100, baos);
        values.put("student_photo", baos.toByteArray());
        try {
            baos.close();
        } catch (IOException e) {
            e.printStackTrace();
        }
        dbw.insert("student", null, values);

        dbw.close();

        SQLiteDatabase dbr = helper.getReadableDatabase();
        String[] columns = {"student_photo", "_id", "student_name",
                "student_birth", "student_phone"};
        Cursor cur = dbr.query("student", columns, null, null, null, null, null);
        int[] li = {R.id.id_li_photo, R.id.id_li_id, R.id.id_li_name,
                R.id.id_li_birth, R.id.id_li_phone};
        cur.moveToNext();
        MySimpleCursorAdapter scad = new MySimpleCursorAdapter(MainActivity.this,
                R.layout.list_item, cur, columns, li);
        lv.setAdapter(scad);

    }
}
```

重点看一下对按钮点击事件的响应方法 onClick()。在 onClick()方法中创建了一个 MyDataBaseHelper 对象,并且调用该对象的 getWritableDatabase()方法,从而获取一个可以写入数据的数据库对象，代码如下：

```
MyDataBaseHelper helper = new MyDataBaseHelper(this);
SQLiteDatabase dbw = helper.getWritableDatabase();
```

然后定义了几个数据变量，代码如下：

```
ContentValues values = new ContentValues();
ByteArrayOutputStream baos;
Bitmap photo;
```

其中的 ContentValues 是要写入数据表的数据的字段名和数据值的 key/value 对。由于 student 表中包含学生头像数据，因此定义了一个 Bitmap 变量和数据流变量，使用这些变量即可向 student 表中插入学生信息记录，代码如下：

```
values.put("student_name", "张大卫");
values.put("student_birth", "2011-01-01");
values.put("student_phone", "13800138000");
photo = BitmapFactory.decodeResource(this.getResources(),
R.drawable.png0010);
baos = new ByteArrayOutputStream();
photo.compress(CompressFormat.PNG, 100, baos);
values.put("student_photo", baos.toByteArray());
try {
    baos.close();
} catch (IOException e) {
    e.printStackTrace();
}
dbw.insert("student", null, values);
```

通过上面的代码，即可向 student 表中插入一条学生信息记录。类似地，我们还需要向 student 表中插入另外两条学生信息记录。在插入学生信息记录后，通过以下代码将 student 表中的记录全部读取出来，并且使用 ListView 组件将其显示出来。

```
SQLiteDatabase dbr = helper.getReadableDatabase();
String[] columns = {"student_photo", "_id", "student_name",
                    "student_birth", "student_phone"};
Cursor cur = dbr.query("student", columns, null, null, null, null, null);
int[] li = {R.id.id_li_photo, R.id.id_li_id, R.id.id_li_name,
            R.id.id_li_birth, R.id.id_li_phone};
cur.moveToNext();
MySimpleCursorAdapter scad = new MySimpleCursorAdapter(MainActivity.this,
                    R.layout.list_item, cur, columns, li);
lv.setAdapter(scad);
```

运行 Ch1103 应用程序，即可显示图 11-7 所示的界面。

现在有一个问题：我们创建的 SQLite 数据库在手机中的什么位置？首先需要知道的是，我们在手机中安装的应用程序，都会安装在手机内部文件系统的/data/data 目录下。为了更加直观地看到这一点，打开 Device File Explorer 面板，即可看到手机内部文件系统。

在手机内部文件系统中展开/data/data 目录，可以看到已经安装的 com.example. ch1103 应用程序目录，展开该目录，在 databases 目录下，可以看到我们创建的 Teach.db 数据库文件，如图 11-9 所示，可以通过 Device File Explorer 面板将该文件从手机中复制到计算机中。

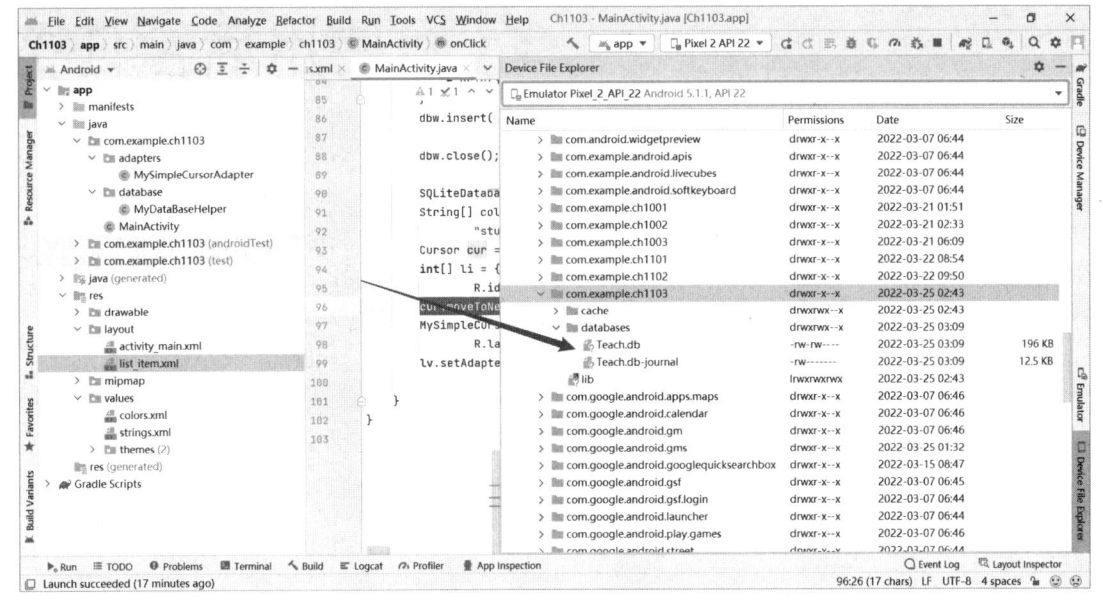

图 11-9　Teach.db 数据库文件

Android 应用程序一旦安装到手机上，Android 平台就会在手机内部文件系统的/data/data 目录下，以该应用程序的包名为名，为该应用程序创建一个目录，这个目录是该应用程序私有的，只有该应用程序和 root 用户才可以读/写这个目录及其子目录。

第 12 章

使用后台任务

在 Android 平台上，Activity 运行在称为 UI 线程的主线程中，并且 Android 平台对 Activity 的响应时间有严格的要求：对每个用户操作的响应时长不能超过规定的时间长度，否则 Android 平台会发生异常，并且会影响用户的使用体验。因此，在 Android 平台上，执行时间较长的任务和执行时间不确定的任务，如网络通信，都必须放在后台执行。

在后台执行任务，完全可以使用 Java 的线程机制进行。在 Android 应用程序中，在后台执行任务是非常普遍的，Android SDK 也提供了相应的机制，使在后台执行任务变得更加容易实现，包括使用 AsyncTask 工具类。本章主要介绍如何在后台执行任务。

12.1 使用 Java 线程执行后台任务

对于一些执行时间较长的任务，可以使用 Java 的线程——Thread 类执行。下面举例说明如何使用 Java 的 Thread 类执行后台任务。

该案例应用程序可以使用一个 TextView 组件实时显示当前系统的日期和时间，运行效果如图 12-1 所示。点击"停止"按钮，可以停止实时显示当前系统的日期和时间，并且按钮上的文字会变为"启动"，再次点击该按钮，可以再次实时显示当前系统的日期和时间。

图 12-1　使用 Java 线程执行后台任务的案例应用程序的运行效果

第 12 章 使用后台任务

下面构建该案例应用程序。在 Android Studio 中新建一个名为 Ch1201 的 Android 应用程序工程。修改布局文件 res/layout/activity_main.xml 中的代码，使其中包含一个 TextView 组件和一个 Button 组件，修改后的代码如下：

```xml
<RelativeLayout xmlns:android="http://schemas.android.com/apk/res/android"
    android:layout_width="match_parent"
    android:layout_height="match_parent">

    <TextView
        android:id="@+id/id_textview"
        android:layout_width="wrap_content"
        android:layout_height="wrap_content"
        android:layout_centerInParent="true"
        android:textSize="24sp" />

    <Button
        android:id="@+id/id_button"
        android:layout_width="match_parent"
        android:layout_height="wrap_content"
        android:layout_alignParentBottom="true"
        android:text="@string/text_button_stop"
        />

</RelativeLayout>
```

修改 res/values/strings.xml 文件中的代码，在其中定义布局文件中引用的字符串资源，修改后的代码如下：

```xml
<resources>
    <string name="app_name">Ch1201</string>

    <string name="text_button_start">启动</string>
    <string name="text_button_stop">停止</string>
</resources>
```

修改 MainActivity.java 文件中的代码，修改后的代码如下：

```java
package com.example.ch1201;

import androidx.appcompat.app.AppCompatActivity;

import android.os.Bundle;
import android.os.Handler;
import android.view.View;
import android.widget.Button;
import android.widget.TextView;
```

```java
import java.text.SimpleDateFormat;
import java.util.Date;
import java.util.Locale;

public class MainActivity extends AppCompatActivity implements View.OnClickListener {
    private TextView tv;
    private Button btn;

    private boolean started;
    private Handler handler;

    private Date d;
    private SimpleDateFormat sdf;

    @Override
    protected void onCreate(Bundle savedInstanceState) {
        super.onCreate(savedInstanceState);
        setContentView(R.layout.activity_main);

        started = true;
        d = new Date();

        tv = (TextView)this.findViewById(R.id.id_textview);
        sdf = new SimpleDateFormat("yyyy-MM-dd HH:mm:ss", Locale.CHINA);
        String ds = sdf.format(d);
        tv.setText(ds);

        btn = (Button)this.findViewById(R.id.id_button);
        btn.setOnClickListener(this);

        handler = new Handler();
        Thread t = new Thread(new MyTimer());
        t.start();
    }

    private class MyTimer implements Runnable {
        @Override
        public void run() {
            while(started) {
                try {
                    Thread.sleep(1000);
```

```java
            } catch (InterruptedException e) {
                e.printStackTrace();
            }

            handler.post(new Runnable(){
                @Override
                public void run() {
                    d.setTime(System.currentTimeMillis());
                    String ds = sdf.format(d);
                    tv.setText(ds);
                }
            });
        }
    }
}

@Override
public void onClick(View v) {
    Button b = (Button)v;

    if (started == true) {
        started = false;
        b.setText(R.string.text_button_start);
    }
    else {
        started = true;
        Thread t = new Thread(new MyTimer());
        t.start();
        b.setText(R.string.text_button_stop);
    }
}
```

在 MainActivity 类中定义了一个 started 变量，用于表示是否实时显示当前系统的日期和时间，还定义了一个类型为 Handler 的变量 handler，该变量的作用是什么呢？这需要从 Android 中 Activity 的工作机制说起。

Android 中的每个 Activity 都运行在自己独立的线程中，这个线程通常称为 UI 线程，Activity 通过 UI 线程与用户进行交互。为了保证界面交互的实时性、不影响用户体验，Android 规定，Activity 完成交互的时间长度不能超过规定的时间长度（10 秒），如果超过规定的时间长度，那么 Android 会发生异常。Android 规定，只能由 UI 线程对 Activity 中显示的信息进行更新，其他线程都不能更新 Activity 中显示的信息，否则会发生异常。为了让非 UI 线程也能修改 Activity 中的信息，Android 为每个 Activity 都提供了一个默认的消息队列，需要修改

Activity 中显示信息的线程通过这个消息队列向 Activity 发送消息，进而由 Activity 的 UI 线程修改 Activity 中显示的信息，Handler 类就是为此设计的。

在 Handler 类的 onCreate()回调方法中，首先显示主界面，然后获取主界面中 TextView 组件和 Button 组件的引用，在 TextView 组件中显示当前系统的日期和时间，再设置对按钮点击事件的响应方法，接着获取消息队列 Handler 对象 handler，用于在非 UI 线程与 UI 线程之间进行通信，最后启动一个后台线程，使其以 1 秒的时间间隔更新界面中显示的日期和时间。后台线程的代码如下：

```java
private class MyTimer implements Runnable {
    @Override
    public void run() {
        while(started) {
            try {
                Thread.sleep(1000);
            } catch (InterruptedException e) {
                e.printStackTrace();
            }

            handler.post(new Runnable(){
                @Override
                public void run() {
                    d.setTime(System.currentTimeMillis());
                    String ds = sdf.format(d);
                    tv.setText(ds);
                }
            });
        }
    }
}
```

在后台线程中，首先判断 started 变量的值是否为 true，若为 true，则表示要修改主界面中显示的日期和时间。因此，每隔 1 秒，都通过 handler 对象向 UI 线程发送一条消息，该消息有点特殊，它要求主界面 UI 线程执行一个特定的方法，在该方法中修改在主界面中显示的当前系统的日期和时间。

对按钮点击事件的响应方法比较简单，只需启动或停止执行后台线程。将 started 变量的值设置为 false，即可停止执行后台线程，当需要再次启动后台线程时，重新创建一个后台线程即可。运行 Ch1201 应用程序，即可得到图 12-1 所示的运行效果。

12.2 同步练习一

读者可以将 12.1 节中的 Ch1201 应用程序复制到自己的开发环境中，并且进行以下修改：每到整点，如早上 8:00、晚上 10:00，都会自动播放一段简短的音乐。

> **提示**
>
> 每到整点，都需要启动一个后台线程，用于播放音乐。

12.3 使用 AsyncTask 工具类执行后台任务

Android 为了便于执行后台任务，提供了 AsyncTask 工具类。使用 AsyncTask 工具类，可以使应用程序在后台执行任务，并且将后台任务的状态或结果显示在主界面 UI 线程中。

使用 AsyncTask 工具类，需要派生 AsyncTask 工具类的一个子类，并且重写 AsyncTask 工具类中的以下 4 个方法。

- protected void onPreExecute()。该方法主要用于在执行后台任务前，执行所需的操作，如进行初始化操作。在该方法中，可以直接将信息显示在主界面中。由于该方法是在主界面 UI 线程中执行的，因此该方法要简短、高效。
- protected Result doInBackground(Params... params)。该方法主要用于执行后台任务。该方法在一个新线程中执行，在使用 execute()方法启动异步任务时，会将参数传递给该方法。在该方法中，通过调用 publishProgress()方法，将任务执行过程中的状态信息传递给 UI 线程，并且将该方法的返回值传递给 UI 线程。
- protected void onProgressUpdate(Progress... values)。在执行后台任务的过程中，可以使用该方法将后台任务执行的中间状态传递给 UI 线程。
- protected void onPostExecute(Result result)。在后台任务执行完毕后，可以使用该方法将后台任务的执行效果通过 UI 线程显示在主界面中。

AsyncTask 工具类支持 3 个泛型的类参数，因此，在派生 AsyncTask 工具类的子类时，需要指出 3 个数据类型：第一个数据类型是传递给 doInBackground()方法的 Params 数据类型，第二个数据类型是传递给 onProgressUpdate()方法的 Progress 数据类型，第三个数据类型是传递给 onPostExecute()方法的 Result 数据类型。对于不需要使用的参数，可以将其对应的 AsyncTask 泛型的类参数设置为 Void 类型。

下面举例说明如何使用 AsyncTask 工具类执行后台任务：使用 AsyncTask 工具类实现 12.1 节中的 Ch1201 应用程序，本案例应用程序运行效果的首界面如图 12-2 所示。点击"停止"按钮，显示的界面如图 12-3 所示。与 Ch1201 应用程序界面不同的是，在本案例应用程序中，我们会使用 Toast 组件显示一条简短的信息，用于告知用户当前日期和时间的状态。

下面构建该案例应用程序。在 Android Studio 中新建一个名为 Ch1202 的 Android 应用程序工程。修改布局文件 res/layout/activity_main.xml 中的代码，使其中包含一个 TextView 组件和一个 Button 组件，修改后的代码如下：

```
<RelativeLayout xmlns:android="http://schemas.android.com/apk/res/android"
    android:layout_width="match_parent"
    android:layout_height="match_parent">

    <TextView
        android:id="@+id/id_textview"
        android:layout_width="wrap_content"
```

```xml
        android:layout_height="wrap_content"
        android:layout_centerInParent="true"
        android:textSize="24sp" />

    <Button
        android:id="@+id/id_button"
        android:layout_width="match_parent"
        android:layout_height="wrap_content"
        android:layout_alignParentBottom="true"
        android:text="@string/text_button_stop"
        />

</RelativeLayout>
```

图 12-2 使用 AsyncTask 工具类显示日期和时间的案例应用程序运行效果的首界面

图 12-3 点击"停止"按钮后显示的界面

修改 res/values/strings.xml 文件中的代码，在其中定义布局文件中引用的字符串资源，修改后的代码如下：

```xml
<resources>
    <string name="app_name">Ch1202</string>

    <string name="text_button_start">启动</string>
    <string name="text_button_stop">停止</string>

</resources>
```

修改 MainActivity.java 文件中的代码，修改后的代码如下：

```java
package com.example.ch1202;

import androidx.appcompat.app.AppCompatActivity;

import android.os.AsyncTask;
import android.os.Bundle;
import android.view.View;
import android.widget.Button;
import android.widget.TextView;
import android.widget.Toast;

import java.text.SimpleDateFormat;
import java.util.Date;
import java.util.Locale;
import java.util.concurrent.atomic.AtomicBoolean;

public class MainActivity extends AppCompatActivity implements View.OnClickListener {
    private TextView tv;
    private Button btn;

    private AtomicBoolean started = new AtomicBoolean();
    private Date d;
    private SimpleDateFormat sdf;

    private MyAsyncTask mat;

    @Override
    protected void onCreate(Bundle savedInstanceState) {
        super.onCreate(savedInstanceState);
        setContentView(R.layout.activity_main);

        started.set(false);
        d = new Date();

        tv = (TextView)this.findViewById(R.id.id_textview);
        sdf = new SimpleDateFormat("yyyy-MM-dd HH:mm:ss", Locale.CHINA);
        String ds = sdf.format(d);
        tv.setText(ds);

        btn = (Button)this.findViewById(R.id.id_button);
        btn.setOnClickListener(this);
    }
```

```java
    @Override
    protected void onResume() {
        super.onResume();

        if (started.get() == false) {
            started.set(true);
            mat = new MyAsyncTask();
            mat.execute();
        }
    }

    @Override
    protected void onPause() {
        super.onPause();

        if (started.get() == true) {
            started.set(false);
        }
    }

    @Override
    public void onClick(View v) {
        if (started.get() == true) {
            started.set(false);
            btn.setText(R.string.text_button_start);
        }
        else {
            started.set(true);
            mat = new MyAsyncTask();
            mat.execute();
            btn.setText(R.string.text_button_stop);
        }
    }

    private class MyAsyncTask extends AsyncTask<Void, Void, Void> {
        @Override
        protected void onPreExecute() {
            Toast.makeText(MainActivity.this, "开始实时显示时间",
                    Toast.LENGTH_SHORT).show();
        }

        @Override
        protected Void doInBackground(Void... params) {
```

```java
            while(started.get() == true) {
                try {
                    Thread.sleep(1000);
                } catch (InterruptedException e) {
                    e.printStackTrace();
                    return null;
                }
                publishProgress();
            }
            return null;
        }

        @Override
        protected void onProgressUpdate (Void… values) {
            d.setTime(System.currentTimeMillis());
            String ds = sdf.format(d);
            tv.setText(ds);
        }

        @Override
        protected void onPostExecute (Void result) {
            Toast.makeText(MainActivity.this, "停止实时显示时间",
                    Toast.LENGTH_SHORT).show();
        }
    }
}
```

在 MainActivity 类的 onCreate() 回调方法中，获取界面中组件的引用，并且设置对按钮点击事件的响应接口。

在 MyAsyncTask 类中，由于 4 个需要重写的方法中均没有参数，因此在 AsyncTask 工具类的泛型类参数中，使用以下代码表示 MyAsyncTask 类的 doInBackground() 方法、onProgressUpdate() 方法和 onPostExecute() 方法，这 3 个方法都不需要参数（Void 就是"无"的意思）。

```
AsyncTask<Void, Void, Void>
```

下面看一下 onPreExecute() 方法的实现，代码如下：

```java
        @Override
        protected void onPreExecute() {
            Toast.makeText(MainActivity.this, "开始实时显示时间",
                                    Toast.LENGTH_SHORT).show();
        }
```

由于 onPreExecute() 方法是在 UI 线程中执行的，因此在该方法中可以放心地对界面中显示的信息进行修改，并且使用 Toast 组件显示一条简短的信息。

下面看一下 doInBackground() 方法的实现，代码如下：

```
        @Override
        protected Void doInBackground(Void… params) {
            while(started.get() == true) {
                try {
                    Thread.sleep(1000);
                } catch (InterruptedException e) {
                    e.printStackTrace();
                    return null;
                }
                publishProgress();
            }
            return null;
        }
```

在 doInBackground()方法中，当 started 变量的值为 true 时，应用程序每隔 1 秒调用一次 publishProgress()方法，进而调用 onProgressUpdate()方法，用于修改界面中显示的信息。需要注意的是，onProgressUpdate()方法也是在 UI 线程中执行的，因此可以放心地修改界面中显示的信息，代码如下：

```
        @Override
        protected void onProgressUpdate (Void… values) {
            d.setTime(System.currentTimeMillis());
            String ds = sdf.format(d);
            tv.setText(ds);
        }
```

最后，当 started 变量的值被设置为 false 时，会导致 doInBackground()方法停止执行，系统会调用 onPostExecute()方法，并且再次使用 Toast 组件显示一条简短的信息，代码如下：

```
        @Override
        protected void onPostExecute (Void result) {
            Toast.makeText(MainActivity.this, "停止实时显示时间",
                                        Toast.LENGTH_SHORT).show();
        }
```

为了启动并运行日期和时间线程，在 Activity 类的 onResume()回调方法中创建了 MyAsyncTask 对象，并且调用该对象的 execute()方法，用于启动并运行后台线程，进而修改界面中显示的当前日期和时间。

在停止实时显示时间时，需要停止后台线程的执行。因此，在 Activity 类的 onPause()回调方法中，我们将 started 变量的值设置为 false，从而停止后台线程的执行，进而达到停止实时显示日期和时间的目的。

由于在 UI 线程及 MyAsyncTask 线程中均需要访问 started 变量，因此为了保证该变量的数据完整性，我们使用了 java.util.concurrent.atomic.AtomicBoolean 数据类型的变量。将 started 变量定义为 AtomicBoolean 数据类型的变量，即可使用 Java JDK 提供的并发控制机制，保证 started 变量在被多个线程使用时的数据完整性。类似地，对于需要在多线程中使用的原始数据类型，也可以使用 java.util.concurrent.atomic 包中的数据类型，用于保证数据的完整性。

运行 Ch1202 应用程序，即可得到图 12-2 所示的运行效果。

12.4 使用 Service 完成后台任务

Service 是 Android 中的组件之一，它没有 UI 接口。使用 Service 可以完成一些需要长时间在后台执行的任务。Android 中的其他组件，如 Activity，都可以启动并运行一个 Service。Service 一旦被启动，就会持续地在后台运行，直到被停止。Android 提供两种类型的 Service，分别为启动式服务和绑定式服务，本节主要介绍启动式服务，对于绑定式服务及 IPC（进程间通信），读者可以参考 Android 帮助文档。

与 Activity 一样，Service 也具有固有的生命周期，可以通过重写生命周期回调方法实现对 Service 的控制。Service 的生命周期如图 12-4 所示。

根据图 12-4 可知，当应用程序的组件调用 startService() 方法时，会启动并运行某个指定的 Service，如果该 Service 之前没有运行过，那么 Android 平台会调用该 Service 的 onCreate() 回调方法初始化该 Service，然后调用该 Service 的 onStartCommand() 方法，使该 Service 进入活动状态；如果该 Service 之前已经被其他的某个应用程序组件启动了，那么 Android 平台会直接调用该 Service 的 onStartCommand() 方法，使该 Service 进入活动状态。当应用程序组件调用 stopService() 方法停止运行 Service 或 Service 自己调用 stopSelf() 方法停止运行时，Android 平台会调用该 Service 的 onDestroy() 回调方法，用于销毁该 Service，结束该 Service 的生命周期。

Service 是后台服务，在启动 Service 的应用程序组件被停止后，Service 仍然可以运行。例如，一个 Activity 启动并运行了某个 Service，但是，在该 Service 被启动后，启动 Service 的 Activity 被用户关闭了。虽然 Activity 被关闭了，但是被 Activity 启动的 Service 仍然可以继续运行。

需要注意的是，虽然 Service 是后台服务，但是 Service 是运行在应用程序的 UI 线程中的，因此在 Service 的各个回调方法中只能进行一些简短的工作，需要长时间执行的任务应该交给其他线程执行。

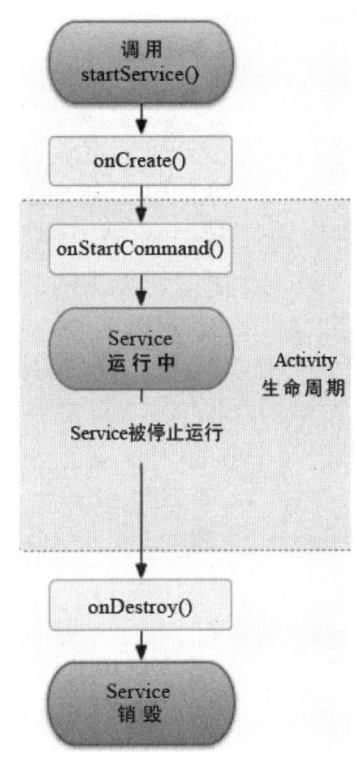

图 12-4 Service 的生命周期

下面举例说明 Service 的使用方法：使用 Service 在后台播放音乐。在第 10 章的相关内容中，音频是在 Activity 中播放的，因此，在关闭播放音频的 Activity 后，系统会停止播放音频。现在对该应用程序进行以下修改：使用 Service 播放音乐，使用 Activity 启动该 Service，在关闭这个 Activity 后，Service 仍然可以在后台播放音乐。该案例应用程序的运行效果如图 12-5 所示。点击"播放 1 号音乐"或"播放 2 号音乐"按钮，该案例应用程序会启动一个 Service，用于播放相应的音乐，同时在系统的通知栏中显示一个播放图标，如图 12-6 所示。在关闭 Activity 后，不会停止播放音乐。通过下拉通知栏，再次启动 Activity，对音乐播放进行控制。

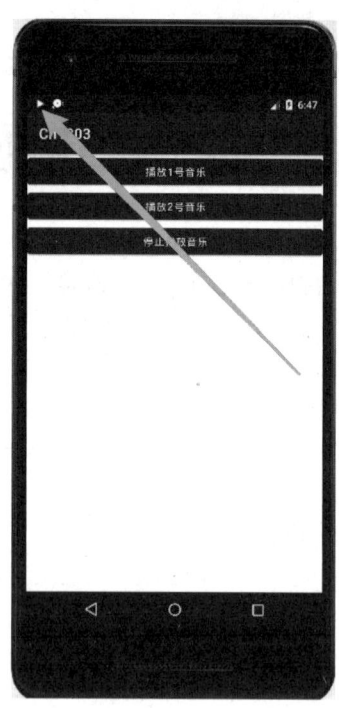

图 12-5 使用 Service 播放音乐的案例应用程序的运行效果　　图 12-6 在系统的通知栏中显示一个播放图标

现在构建该案例应用程序。在 Android Studio 中新建一个名为 Ch1203 的 Android 应用程序工程。修改布局文件 res/layout/activity_main.xml 中的代码，修改后的代码如下：

```xml
<RelativeLayout xmlns:android="http://schemas.android.com/apk/res/android"
    android:layout_width="match_parent"
    android:layout_height="match_parent">

    <Button
        android:id="@+id/id_button_1"
        android:layout_width="match_parent"
        android:layout_height="wrap_content"
        android:layout_alignParentTop="true"
        android:text="@string/text_button_1" />

    <Button
        android:id="@+id/id_button_2"
        android:layout_width="match_parent"
        android:layout_height="wrap_content"
        android:layout_below="@id/id_button_1"
        android:text="@string/text_button_2" />

    <Button
        android:id="@+id/id_button_3"
        android:layout_width="match_parent"
```

```xml
            android:layout_height="wrap_content"
            android:layout_below="@id/id_button_2"
            android:text="@string/text_button_3" />

</RelativeLayout>
```

修改 res/values/strings.xml 文件中的代码，在其中定义布局文件中引用的字符串资源，修改后的代码如下：

```xml
<?xml version="1.0" encoding="utf-8"?>
<resources>
    <string name="app_name">Ch1203</string>

    <string name="text_button_1">播放 1 号音乐</string>
    <string name="text_button_2">播放 2 号音乐</string>
    <string name="text_button_3">停止播放音乐</string>

</resources>
```

现在编写用于播放音乐的 Service 程序的代码。为了便于管理，在 src 目录下新建一个名为 com.example.ch1203.service 的包，在该包中新建一个名为 MyService 的 Java 类文件。修改 MyService.java 文件中的代码，修改后的代码如下：

```java
package com.example.ch1203.service;

import android.app.Notification;
import android.app.NotificationManager;
import android.app.PendingIntent;
import android.app.Service;
import android.content.Intent;
import android.media.AudioManager;
import android.media.MediaPlayer;
import android.os.Environment;
import android.os.IBinder;
import android.widget.Toast;

import com.example.ch1203.MainActivity;

import java.io.IOException;

public class MyService extends Service {
    private MediaPlayer mp;

    private NotificationManager mNM;
    private int mNOTIFICATION = (int)System.currentTimeMillis();
```

```java
@Override
public void onCreate() {
    mp = null;
    mNM = (NotificationManager)getSystemService(NOTIFICATION_SERVICE);
    showNotification();

    System.out.println("onCreate called");
}

@Override
public int onStartCommand(Intent intent, int flags, int startId) {
    System.out.println("onStartCommand called");

    if (mp != null) {
        if (mp.isPlaying() == true) {
            mp.stop();
        }
        mp.release();
    }

    mp = new MediaPlayer();

    mp.setAudioStreamType(AudioManager.STREAM_MUSIC);
    mp.setOnPreparedListener(new MediaPlayer.OnPreparedListener() {
        @Override
        public void onPrepared(MediaPlayer mp) {
            mp.start();
        }
    });
    mp.setOnErrorListener(new MediaPlayer.OnErrorListener() {
        @Override
        public boolean onError(MediaPlayer mp, int what, int extra) {
            Toast.makeText(MyService.this, "无法播放该文件",
                    Toast.LENGTH_LONG).show();
            return false;
        }
    });
    mp.setOnCompletionListener(new MediaPlayer.OnCompletionListener() {
        @Override
        public void onCompletion(MediaPlayer mp) {
            MyService.this.stopSelf();
        }
    });
```

```java
        try {
            String music = intent.getStringExtra("music");

            String sd_card_path =
                    Environment.getExternalStorageDirectory().getAbsolutePath();
            mp.setDataSource(sd_card_path + music);
            mp.prepareAsync();
        } catch (IllegalArgumentException e) {
            e.printStackTrace();
        } catch (SecurityException e) {
            e.printStackTrace();
        } catch (IllegalStateException e) {
            e.printStackTrace();
        } catch (IOException e) {
            e.printStackTrace();
        }

        return START_STICKY;
    }

    @Override
    public IBinder onBind(Intent intent) {
        // 由于不提供绑定式服务，因此直接返回 null
        return null;
    }

    @Override
    public void onDestroy() {
        if (mp != null) {
            if (mp.isPlaying() == true) {
                mp.stop();
            }
            mp.release();
        }

        mNM.cancel(mNOTIFICATION);
        System.out.println("onDestroy called");
    }

    @SuppressWarnings("deprecation")
    private void showNotification() {
        PendingIntent contentIntent = PendingIntent.getActivity(this, 0,
```

```
                new Intent(this, MainActivity.class), 0);

        CharSequence text = "Playing Music";
        Notification.Builder builder = new Notification.Builder(this);
        builder.setContentTitle(text);
        builder.setContentText(text + "...");
        builder.setSmallIcon(android.R.drawable.ic_media_play);
        builder.setContentIntent(contentIntent);//执行intent
        Notification notification = builder.getNotification();
        mNM.notify(mNOTIFICATION, notification);
    }
}
```

在 MyService 类的 onCreate()回调方法中，先对相关变量进行初始化，再在系统的通知栏中显示一条通知，并且在该通知中显示一个播放图标。在 onStartCommand()回调方法中，先停止之前播放的音乐（如果之前正在播放的话），再设置音乐播放的相关接口，注意 OnCompletionListener 接口，当该接口中的响应方法被调用时，表示指定的音乐已经播放完毕，因此，在该接口中的响应方法中停止被启动的 Service，代码如下：

```
        mp.setOnCompletionListener(new OnCompletionListener() {
            @Override
            public void onCompletion(MediaPlayer mp) {
                MyService.this.stopSelf();
            }
        });
```

onStartCommand()回调方法的返回值及其含义如下。
- START_NOT_STICKY。当系统资源紧缺时，在 onStartCommand()回调方法返回数据后，如果系统结束了这个 Service，则不需要重新创建和启动这个 Service。
- START_STICKY。当系统资源紧缺时，在 onStartCommand()回调方法返回数据后，如果系统结束了这个 Service，则需要重新创建和启动这个 Service，并且再次使用 Intent 对象为 null 的参数调用该 Service 的 onStartCommand()回调方法。
- START_REDELIVER_INTENT。当系统资源紧缺时，在 onStartCommand()回调方法返回数据后，如果系统结束了这个 Service，则需要重新创建和启动这个 Service，并且使用之前调用 onStartCommand()回调方法时的 Intent 参数再次调用该 Service 的 onStartCommand()回调方法。

无论采用何种方式结束 Service，系统都会调用 MyService 类中的 onDestroy()回调方法，用于停止正在播放的音乐，并且取消在通知栏中显示的通知。

修改 MainActivity.java 文件中的代码，修改后的代码如下：

```
package com.example.ch1203;

import androidx.appcompat.app.AppCompatActivity;
```

```java
    import android.content.Intent;
    import android.os.Bundle;
    import android.view.View;
    import android.widget.Button;

    import com.example.ch1203.service.MyService;

    public class MainActivity extends AppCompatActivity implements
View.OnClickListener {

        @Override
        protected void onCreate(Bundle savedInstanceState) {
            super.onCreate(savedInstanceState);
            setContentView(R.layout.activity_main);

            Button btn1 = (Button)this.findViewById(R.id.id_button_1);
            btn1.setOnClickListener(this);
            Button btn2 = (Button)this.findViewById(R.id.id_button_2);
            btn2.setOnClickListener(this);
            Button btn3 = (Button)this.findViewById(R.id.id_button_3);
            btn3.setOnClickListener(this);
        }

        @Override
        public void onClick(View v) {
            int id = v.getId();

            if (id == R.id.id_button_1) {
                Intent service = new Intent(this, MyService.class);
                service.putExtra("music", "/Ring01.wav");
                this.startService(service);
            }
            else if (id == R.id.id_button_2) {
                Intent service = new Intent(this, MyService.class);
                service.putExtra("music", "/Ring02.wav");
                this.startService(service);
            }
            else {
                System.out.println("Stopping Service");
                Intent service = new Intent(this, MyService.class);
                this.stopService(service);
```

```
        }
    }
}
```

在 MainActivity 类中显示主界面,并且设置对按钮点击事件的响应方法 onClick()。在 onClick()方法中,当点击第一个或第二个按钮时,创建一个用于启动 Service 的 Intent 对象,并且传递一个需要播放的音乐文件名,然后调用 startService()方法,用于启动并运行指定的 Service;当点击第三个按钮时,调用 stopService()方法,用于结束指定的 Service。

在运行 Ch1203 应用程序前,需要在 SD 卡中存储两个音乐文件:/Ring01.wav 和/Ring02.wav。

修改 AndroidManifest.xml 文件中的代码,在其中声明 Service 程序 MyService,以及为读取 SD 卡授权,修改后的代码如下:

```xml
<?xml version="1.0" encoding="utf-8"?>
<manifest xmlns:android="http://schemas.android.com/apk/res/android"
    package="com.example.ch1203">

    <uses-permission android:name="android.permission.READ_EXTERNAL_STORAGE"/>

    <application
        android:allowBackup="true"
        android:icon="@mipmap/ic_launcher"
        android:label="@string/app_name"
        android:roundIcon="@mipmap/ic_launcher_round"
        android:supportsRtl="true"
        android:theme="@style/Theme.Ch1203">
        <activity
            android:name=".MainActivity"
            android:exported="true">
            <intent-filter>
                <action android:name="android.intent.action.MAIN" />

                <category android:name="android.intent.category.LAUNCHER" />
            </intent-filter>
        </activity>

        <service android:name=".service.MyService"/>

    </application>

</manifest>
```

以下代码主要用于声明需要使用的 Service 程序 MyService。

```xml
<service android:name="com.example.ch1203.service.MyService"/>
```

运行 Ch1203 应用程序,即可得到图 12-5 所示的运行效果。

12.5 同步练习二

Android 提供了一个实用的 Service 子类——IntentService 类。使用 IntentService 类可以简化 Service 程序的编写内容。自学 IntentService 类，并且使用 IntentService 类完成实时显示日期和时间的应用程序。

> **提示**
>
> 重写 IntentService 类中的 onHandleIntent()方法，通过延时操作，修改界面中日期和时间的显示状态。

第 13 章

使用网络

在互联网时代，几乎没有 Android 应用程序是独立于网络之外存在的，因此，开发基于网络的应用程序是必要的。本章主要介绍在 Android 应用程序中如何使用网络，包括使用 ConnectivityManager 管理网络状态、使用 HttpURLConnection 访问网络、使用 OkHttp 访问网络、使用 Multipart 传递请求数据到服务器端、使用 JSON（JavaScript Object Notation）格式的数据与服务器端进行通信。

13.1 使用 ConnectivityManager 管理网络状态

在使用网络进行数据通信前，需要获取网络的状态。例如，当前网络是否开启，网络连接方式是哪种（Wi-Fi 连接、GPRS 连接、UMTS 连接），等等。使用 Android SDK 提供的 ConnectivityManager 类可以获取网络的状态。通过以下代码可以获取一个 ConnectivityManager 对象。

```
ConnectivityManager cm = (ConnectivityManager)
    Context.getSystemService(Context.CONNECTIVITY_SERVICE);
```

使用 ConnectivityManager 类中的方法和属性可以获取网络状态和监听网络状态的变化。ConnectivityManager 类中常用的方法如下。

- public NetworkInfo getActiveNetworkInfo()。获取当前活动的网络信息。如果当前没有活动的网络，则返回 null。
- public NetworkInfo[] getAllNetworkInfo()。获取系统支持的所有网络信息。

如果要获取网络状态或监听网络状态的变化，那么应用程序需要在 AndroidManifest.xml 文件中申请 android.permission.ACCESS_NETWORK_STATE 权限。

下面举例说明 ConnectivityManager 类的使用方法。首先检查当前是否有活动的网络，如果有，则打印该活动网络的相关信息。该案例应用程序的运行效果如图 13-1 所示。由于该案例应用程序是在模拟器中运行的，而模拟器中的 mobile 数据网络是开启的，因此，该案例应用程序在 TextView 组件中显示 mobile 数据网络是可用的。

下面构建该案例应用程序。在 Android Studio 中新建一个

图 13-1 检查当前是否有活动网络的案例应用程序的运行效果

名为 Ch1301 的 Android 应用程序工程。修改布局文件 res/layout/activity_main.xml 中的代码，为其中的 TextView 组件添加一个 id 属性，修改后的代码如下：

```xml
<RelativeLayout xmlns:android="http://schemas.android.com/apk/res/android"
    android:layout_width="match_parent"
    android:layout_height="match_parent">

    <TextView
        android:id="@+id/id_textview"
        android:layout_width="wrap_content"
        android:layout_height="wrap_content"
        android:text="" />

</RelativeLayout>
```

修改 MainActivity.java 文件中的代码，修改后的代码如下：

```java
package com.example.ch1301;

import androidx.appcompat.app.AppCompatActivity;

import android.net.ConnectivityManager;
import android.net.NetworkInfo;
import android.os.Bundle;
import android.widget.TextView;

public class MainActivity extends AppCompatActivity {

    @Override
    protected void onCreate(Bundle savedInstanceState) {
        super.onCreate(savedInstanceState);
        setContentView(R.layout.activity_main);

        StringBuffer sb = new StringBuffer();

        ConnectivityManager cm = (ConnectivityManager)
                this.getSystemService(MainActivity.CONNECTIVITY_SERVICE);
        NetworkInfo ni = cm.getActiveNetworkInfo();
        if (ni == null) {
            sb.append("当前没有活动网络。");
        }
        else {
            if (ni.isConnected()){
                sb.append(ni.getTypeName()).append("是活动的。");
            }
```

```
            else {
                sb.append(ni.getTypeName()).append("不在服务区。");
            }
        }

        TextView tv = (TextView)this.findViewById(R.id.id_textview);
        tv.setText(sb.toString());
    }
}
```

在 MainActivity 类的 onCreate()回调方法中，首先显示主界面，然后获取 ConnectivityManager 对象 cm，再调用 cm 对象的 getActiveNetworkInfo()方法，用于获取当前活动网络的 NetworkInfo 对象 ni，如果这个方法返回 null，则表示当前没有活动的网络，否则表示当前有活动的网络。为了保证当前网络可以进行数据通信，需要进一步调用 ni 对象的 isConnected()方法，判断当前网络是否可用，并且显示相应的信息。

要运行 Ch1301 应用程序，还需要在 AndroidManifest.xml 文件中添加以下代码，用于对网络进行相应的授权申请。

```
<uses-permission
        android:name="android.permission.ACCESS_NETWORK_STATE"/>
```

运行 Ch1301 应用程序，即可得到图 13-1 所示的运行效果。

13.2 使用 HttpURLConnection 访问网络

使用 HttpURLConnection 访问网络是与服务器端进行基于 HTTP 通信的直接方式。使用 HttpURLConnection 与后台进行通信的过程如下。

（1）使用 URL.openConnection()方法获取一个 HttpURLConnection 对象。
（2）设置请求头的相关参数。
（3）在使用 POST 方法的请求中调用 setDoOutput(true)方法，并且使用 getOutputStream()方法获取输出流，进而利用该输出流输出数据。
（4）处理响应数据。
（5）使用 disconnect()方法关闭网络连接。

HttpURLConnection 支持 HTTP 中规定的所有请求方法，基于上面介绍的过程，可以使用 GET、POST、HEAD、OPTION、DELETE、TRACE 方法向服务器端发送请求，其中，常用的是 GET 方法和 POST 方法，下面介绍如何使用这两种方法发送请求并处理响应数据。

为了便于观察 Android 与服务器端进行通信的过程，编写一个服务器端程序，使用 Servlet 处理来自客户端的 HTTP 请求（GET 请求和 POST 请求）。该服务器端程序主要用于完成以下任务：接收来自客户端的 HTTP 请求，并且根据请求参数 type 和 id 向请求客户端发送指定的图片数据。具体来说就是，客户端使用 GET 方法或 POST 方法向服务器端发送请求，其中的 type 参数表示请求的图片类型，type=1 表示请求蝴蝶图片，type=2 表示请求卡通图片；id 参数表示请求的图片编号。该 Servlet 的代码如下：

```java
package com.ttt.servlet;

import javax.servlet.ServletException;
import javax.servlet.ServletOutputStream;
import javax.servlet.annotation.WebServlet;
import javax.servlet.http.HttpServlet;
import javax.servlet.http.HttpServletRequest;
import javax.servlet.http.HttpServletResponse;
import java.io.FileInputStream;
import java.io.IOException;

@WebServlet("/ImageShower")
public class ImageShower extends HttpServlet {
    private static final long serialVersionUID = 1L;

    protected void doGet(HttpServletRequest request, HttpServletResponse response)
            throws ServletException, IOException {
        String type = request.getParameter("type");
        if ((type == null) || (type.equalsIgnoreCase(""))) {
            type = "1";
        }
        String id = request.getParameter("id");
        if ((id == null) || (id.equalsIgnoreCase(""))) {
            id = "1";
        }

        FileInputStream fis = new FileInputStream(this.getServletContext().
                getRealPath("") + "images/png" + type + id + ".png");
        byte[] b=new byte[fis.available()];
        fis.read(b);
        fis.close();

        response.setContentType("image/png");
        ServletOutputStream op = response.getOutputStream();
        op.write(b);
        op.close();
    }

    protected void doPost(HttpServletRequest request, HttpServletResponse response)
            throws ServletException, IOException {
```

```
            doGet(request, response);
    }
}
```

将 4 张图片存储于 image 目录下,文件名为 pngxx.png,xx 可以为 "11" "12" "21" "22"。4 张图片的效果如图 13-2 所示。

图 13-2　4 张图片的效果

在 Android 客户端使用 HttpURLConnection 获取并显示图片。下面分别介绍使用 GET 方法和 POST 方法获取图片的方法。

13.2.1　使用 HttpURLConnection 的 GET 方法获取图片

使用 HttpURLConnection 的 GET 方法获取服务器端的图片,并且将其显示在手机屏幕上,该案例应用程序的运行效果如图 13-3 所示。

点击 4 个按钮中的任意一个,会在手机屏幕上显示相应的图片。例如,点击 "显示第一张卡通" 按钮,显示的界面如图 13-4 所示。

图 13-3　使用 GET 方法获取服务器端图片的案例应用程序的运行效果

图 13-4　点击 "显示第一张卡通" 按钮后显示的界面

现在构建该案例应用程序。在 Android Studio 中新建一个名为 Ch1302 的 Android 应用程序工程。修改布局文件 res/layout/activity_main.xml 中的代码，修改后的代码如下：

```xml
<LinearLayout xmlns:android="http://schemas.android.com/apk/res/android"
    android:layout_width="match_parent"
    android:layout_height="match_parent"
    android:orientation="vertical">

    <Button
        android:id="@+id/id_btn_1"
        android:layout_width="match_parent"
        android:layout_height="wrap_content"
        android:text="@string/text_btn_1" />

    <Button
        android:id="@+id/id_btn_2"
        android:layout_width="match_parent"
        android:layout_height="wrap_content"
        android:text="@string/text_btn_2" />
    <Button
        android:id="@+id/id_btn_3"
        android:layout_width="match_parent"
        android:layout_height="wrap_content"
        android:text="@string/text_btn_3" />

    <Button
        android:id="@+id/id_btn_4"
        android:layout_width="match_parent"
        android:layout_height="wrap_content"
        android:text="@string/text_btn_4" />

    <ImageView
        android:id="@+id/id_iv"
        android:layout_width="match_parent"
        android:layout_height="match_parent"
        android:scaleType="fitCenter"
        android:contentDescription="@string/hello_world"
        />

</LinearLayout>
```

这个布局文件很简单，主要包含 4 个 Button 组件和一个 ImageView 组件。

修改 res/values/strings.xml 文件中的代码，在其中定义布局文件中引用的字符串资源，修改后的代码如下：

```xml
<resources>
    <string name="app_name">Ch1302</string>

    <string name="text_btn_1">显示第一张蝴蝶</string>
    <string name="text_btn_2">显示第二张蝴蝶</string>
    <string name="text_btn_3">显示第一张卡通</string>
    <string name="text_btn_4">显示第二张卡通</string>

    <string name="hello_world">Hello world!</string>

</resources>
```

修改 MainActivity.java 文件中的代码，修改后的代码如下：

```java
package com.example.ch1302;

import androidx.appcompat.app.AppCompatActivity;

import android.graphics.Bitmap;
import android.graphics.BitmapFactory;
import android.net.ConnectivityManager;
import android.net.NetworkInfo;
import android.os.AsyncTask;
import android.os.Bundle;
import android.view.View;
import android.widget.Button;
import android.widget.ImageView;
import android.widget.Toast;

import java.io.BufferedInputStream;
import java.io.ByteArrayOutputStream;
import java.io.IOException;
import java.net.HttpURLConnection;
import java.net.MalformedURLException;
import java.net.URL;

public class MainActivity extends AppCompatActivity implements View.OnClickListener {
    private ImageView iv;

    @Override
    protected void onCreate(Bundle savedInstanceState) {
        super.onCreate(savedInstanceState);
        setContentView(R.layout.activity_main);
```

```java
        iv = (ImageView) this.findViewById(R.id.id_iv);

        Button btn_1 = (Button) this.findViewById(R.id.id_btn_1);
        btn_1.setOnClickListener(this);
        Button btn_2 = (Button) this.findViewById(R.id.id_btn_2);
        btn_2.setOnClickListener(this);
        Button btn_3 = (Button) this.findViewById(R.id.id_btn_3);
        btn_3.setOnClickListener(this);
        Button btn_4 = (Button) this.findViewById(R.id.id_btn_4);
        btn_4.setOnClickListener(this);
    }

    @Override
    public void onClick(View v) {
        if (checkNetworkState() != true) {
            Toast.makeText(this, "网络没有打开，请打开网络后再试。",
                    Toast.LENGTH_LONG).show();
            return;
        }

        int id = v.getId();
        switch (id) {
            case R.id.id_btn_1:
                downLoadImageAndShow(1, 1);
                break;
            case R.id.id_btn_2:
                downLoadImageAndShow(1, 2);
                break;
            case R.id.id_btn_3:
                downLoadImageAndShow(2, 1);
                break;
            case R.id.id_btn_4:
                downLoadImageAndShow(2, 2);
                break;
        }

    }

    private boolean checkNetworkState() {
        ConnectivityManager cm = (ConnectivityManager) this
                .getSystemService(MainActivity.CONNECTIVITY_SERVICE);
        NetworkInfo ni = cm.getActiveNetworkInfo();
```

```java
            if ((ni == null) || (ni.isConnected() == false)) {
                return false;
            }

            return true;
        }

        private void downLoadImageAndShow(int type, int id) {
            new MyAsyncTask().
                    execute("http://192.168.197.128:8080/an/ImageShower?type=" +
                            type + "&" + "id=" + id);
        }

        private class MyAsyncTask extends AsyncTask<String, Void, Bitmap> {
            @Override
            protected void onPreExecute() {
            }

            @Override
            protected void onProgressUpdate (Void... values) {
            }

            @Override
            protected void onPostExecute (Bitmap bm) {
                iv.setImageBitmap(bm);
            }

            @Override
            protected Bitmap doInBackground(String... params) {
                URL url;
                HttpURLConnection urlConnection = null;
                Bitmap bm = null;

                try {
                    url = new URL(params[0]);
                    urlConnection = (HttpURLConnection)url.openConnection();
                    urlConnection.setRequestMethod("GET");

                    urlConnection.connect();
                    if (urlConnection.getResponseCode() !=
HttpURLConnection.HTTP_OK) {
                        urlConnection.disconnect();
                        return null;
```

```
            }
            byte[] b = new byte[2048];
            ByteArrayOutputStream baos = new ByteArrayOutputStream();

            BufferedInputStream in = new
                BufferedInputStream(urlConnection.getInputStream());
            int len = 0;
            while((len = in.read(b))>0) {
                baos.write(b, 0, len);
            }

            bm = BitmapFactory.decodeByteArray(baos.toByteArray(), 0, baos.size());
            baos.close();
            in.close();

        }catch (MalformedURLException e) {
            return null;
        }catch (IOException e) {
            return null;
        }
        finally {
            urlConnection.disconnect();
        }

        return bm;
    }
}
```

在 MainActivity 类的 onCreate()回调方法中，先显示主界面，再获取相关组件的引用，并且监听对按钮的点击事件。在对按钮点击事件的响应方法 onClick()中，首先判断当前是否有可用的网络，若有，则调用 downLoadImageAndShow()方法下载图片，并且将其显示在 ImageView 组件中。

由于网络操作需要在独立的线程中进行，因此在 downLoadImageAndShow()方法中创建一个 MyAsyncTask 对象（MyAsyncTask 类是 AsyncTask 工具类的子类），用于完成网络操作。在 MyAsyncTask 类的 doInBackground()方法中，首先定义一个指向服务器端 Servlet 的 URL 对象，用于建立网络连接并获取 HttpURLConnection 对象，然后设置请求方法为 GET 方法并发送请求，接下来检查服务器端是否正确处理了客户端的请求，若是，则接收从服务器端 Servlet 发来的图片二进制流，然后使用 BitmapFactory 将其解码为 Bitmap 对象，再将 Bitmap 对象返回给 PostExcute()方法，最后将图片显示在 ImageView 组件中。

要运行 Ch1302 应用程序，还需要在 AndroidManifest.xml 文件中对网络进行相应的授权申请，代码如下：

```xml
<uses-permission android:name="android.permission.ACCESS_NETWORK_STATE"/>
<uses-permission android:name="android.permission.INTERNET"/>
```

运行 Ch1302 应用程序，即可得到图 13-3 所示的运行效果。

13.2.2 使用 HttpURLConnection 的 POST 方法获取图片

使用 HttpURLConnection 的 POST 方法获取服务器端的图片，并且将其显示在手机屏幕上，该应用程序的运行效果与图 13-3 所示的运行效果相同。我们只需修改 downLoadImageAndShow() 方法中的代码，使其使用 POST 方法获取图片数据。修改后的 downLoadImageAndShow() 方法如下：

```java
private void downLoadImageAndShow(int type, int id) {
    new MyAsyncTask(type,id).
            execute("http://192.168.197.128:8080/an/ImageShower");
}

private class MyAsyncTask extends AsyncTask<String, Void, Bitmap> {
    private int type, id;

    public MyAsyncTask(int type, int id) {
        this.type = type;
        this.id = id;
    }

    @Override
    protected void onPreExecute() {
    }

    @Override
    protected void onProgressUpdate (Void... values) {
    }

    @Override
    protected void onPostExecute (Bitmap bm) {
        iv.setImageBitmap(bm);
    }

    @Override
    protected Bitmap doInBackground(String... params) {
```

```java
            URL url;
            HttpURLConnection urlConnection = null;
            Bitmap bm = null;

            try {
                url = new URL(params[0]);
                urlConnection = (HttpURLConnection)url.openConnection();

                urlConnection.setRequestMethod("POST");
                urlConnection.setDoInput(true);
                urlConnection.setDoOutput(true);

                urlConnection.setUseCaches(false);
                urlConnection.setInstanceFollowRedirects(true);
                urlConnection.setChunkedStreamingMode(0);
                urlConnection.setRequestProperty("Content-Type",
                        "application/x-www-form-urlencoded");

                urlConnection.connect();
                DataOutputStream out = new
                        DataOutputStream(urlConnection.getOutputStream());

                String content = "type=" + URLEncoder.encode("" + type, "UTF-8") +
                        "&id=" + URLEncoder.encode("" + id, "UTF-8");
                out.writeBytes(content);
                out.flush();
                out.close();

                if (urlConnection.getResponseCode() !=
HttpURLConnection.HTTP_OK) {
                    urlConnection.disconnect();
                    return null;
                }

                byte[] b = new byte[2048];
                ByteArrayOutputStream baos = new ByteArrayOutputStream();
                BufferedInputStream in = new
                        BufferedInputStream(urlConnection.getInputStream());
                int len = 0;
                while((len = in.read(b))>0) {
                    baos.write(b, 0, len);
                }
```

```
                    bm = BitmapFactory.decodeByteArray(baos.toByteArray(), 0,
baos.size());
                    baos.close();
                    in.close();
                }catch (MalformedURLException e) {
                    return null;
                }catch (IOException e) {
                    return null;
                }
                finally {
                    urlConnection.disconnect();
                }

                return bm;
            }
        }
```

与使用 GET 方法获取图片不同的关键在于以下代码。

```
                    urlConnection.setRequestMethod("POST");
```

上述代码主要用于将访问方法设置为 POST。

以下代码是必需的，用于告诉 HttpURLConnection 对象我们会使用 HTTP 请求体发送请求数据。

```
                    urlConnection.setDoInput(true);
                    urlConnection.setDoOutput(true);
```

以下代码表示为了加快数据传输的速度，设置每个数据块的大小为默认数据块大小。

```
                    urlConnection.setChunkedStreamingMode(0);
```

以下代码表示以表单的形式向服务器端发送请求数据。

```
                    urlConnection.setRequestProperty("Content-Type",
                            "application/x-www-form-urlencoded");
```

使用以下代码获取一个连接到服务器端的数据流，并且向服务器端发送请求数据。需要注意的是，数据体部分采用 UTF-8 编码方案。其他代码与使用 GET 方法获取图片的代码类似，此处不再赘述。

```
                    urlConnection.connect();
                    DataOutputStream out = new
                            DataOutputStream(urlConnection.getOutputStream());

                    String content = "type=" + URLEncoder.encode("" + type, "UTF-8") +
                            "&id=" + URLEncoder.encode("" + id, "UTF-8");
                    out.writeBytes(content);
                    out.flush();
                    out.close();
```

运行修改后的 Ch1302 应用程序，即可得到图 13-3 所示的运行效果。

现在我们还需要回答一个问题：既然使用 GET 方法和 POST 方法都可以进行网络通信，那么在什么情况下使用 GET 方法，在什么情况下使用 POST 方法呢？答案是，当使用 GET 方法进行网络请求时，请求参数附加在 URL 地址的后面。这种方法虽然简单，但是不安全，并且参数长度不能超过 2048 字节；而使用 POST 方法进行网络请求，请求参数是放置在请求体中的，安全且不受请求参数长度的限制。

13.3 同步练习一

仿照 13.2 节中的 Ch1302 应用程序，使用基于 Java 的多线程机制（Thread 机制）实现与 Ch1302 应用程序类似的功能，并且在应用程序界面中使用 GET 方法或 POST 方法获取图片。

13.4 使用 OkHttp 访问网络

由于 Apache 的 HttpClient 非常庞大且复杂，因此从 Android 5.1 开始，Google 的 Android 开发团队不再将 Apache 的 HttpClient 纳入 Android SDK，要使用 HTTP，读者可以使用 13.2 节介绍的 HttpURLConnection 或第三方开源的 HTTP 包。这里，我们使用完善且易用的 OkHttp。

13.4.1 使用 GET 方法进行服务请求

要使用 OkHttp，需要在 OkHttp 官方网站下载最新的 OkHttp 包和 Okio 包，在 Android 应用程序工程的 Project 视图下，将其放置到工程的 app/libs 目录下，如图 13-5 所示。

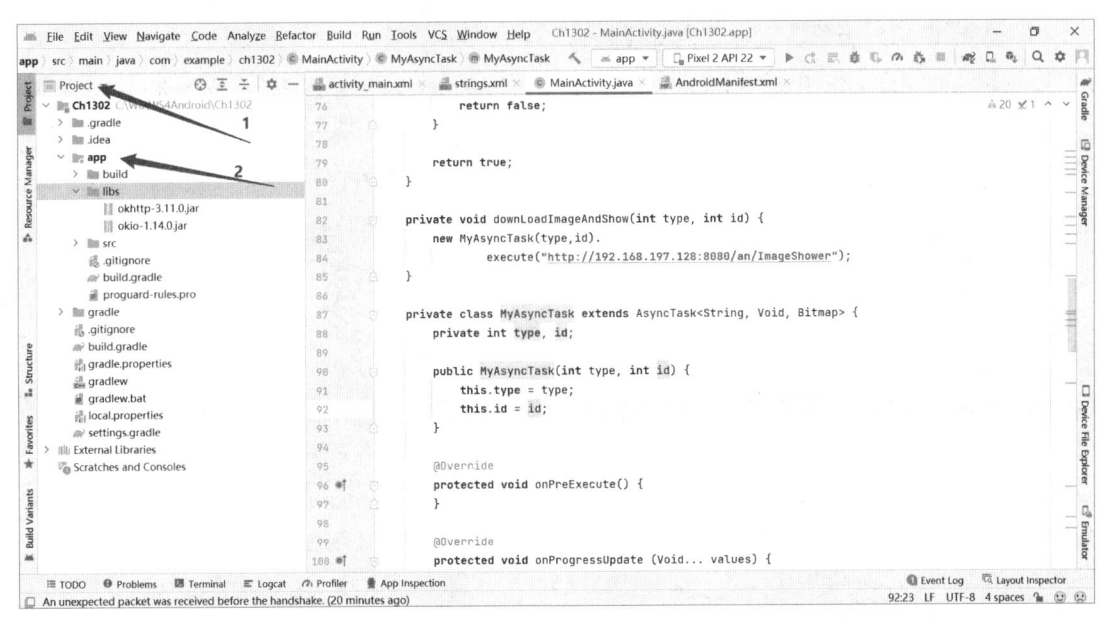

图 13-5 将 OkHttp 包和 Okio 包放置到 Android 应用程序工程的 app/libs 目录下

分别右击这两个包，在弹出的快捷菜单中选择 Add As Library 命令，如图 13-6 所示。
在完成这些准备工作后，就可以使用 OkHttp 了，基本过程如下。

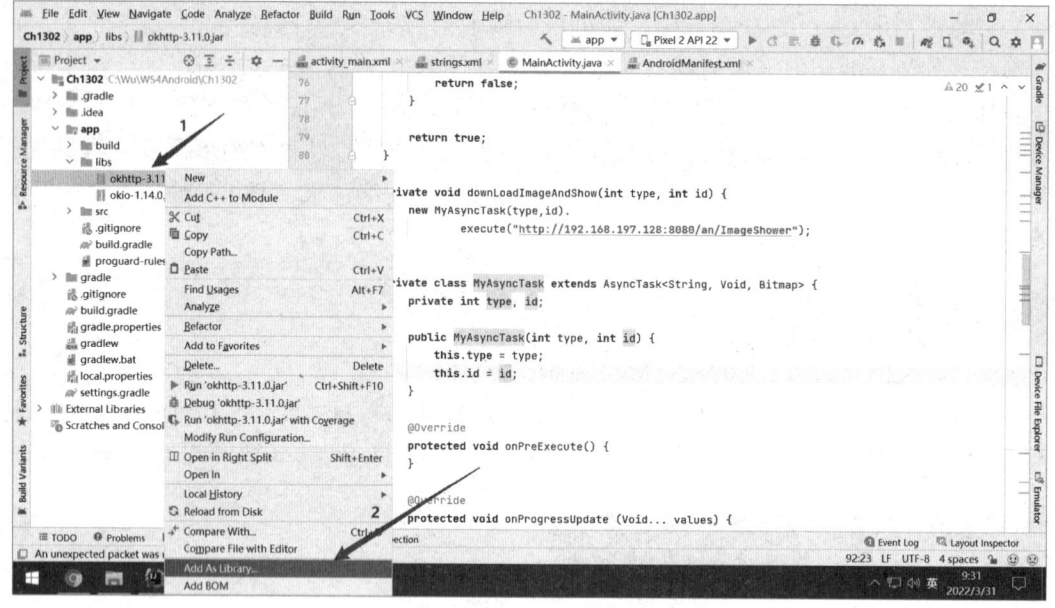

图 13-6　选择 Add as Library 命令

(1) 创建 OkHttpClient 对象，代码如下：

```
private final OkHttpClient client = new OkHttpClient();
```

(2) 创建 HTTP 请求，代码如下：

```
Request request = new Request.Builder()
        .url("http://publicobject.com/helloworld.txt")
        .build();
```

(3) 发送请求到服务器端，代码如下：

```
Response response = client.newCall(request).execute();
```

(4) 处理从服务器端返回的结果，代码如下：

```
if (!response.isSuccessful()) throw new IOException("Unexpected code " + response);
Headers responseHeaders = response.headers();
for (int i = 0; i < responseHeaders.size(); i++) {
    System.out.println(responseHeaders.name(i) + ": " + responseHeaders.value(i));
}
System.out.println(response.body().string());
```

第 (3) 步和第 (4) 步使用同步方式发送请求和接收响应，也可以使用异步方式发送请求和接收响应，代码如下：

```
client.newCall(request).enqueue(new Callback() {
    @Override public void onFailure(Request request, Throwable throwable) {
        throwable.printStackTrace();
    }
```

```
            @Override public void onResponse(Response response) throws IOException {
                if (!response.isSuccessful()) throw new IOException("Unexpected code " + response);
                Headers responseHeaders = response.headers();
                for (int i = 0; i < responseHeaders.size(); i++) {
                    System.out.println(responseHeaders.name(i) + ": " +
                                        responseHeaders.value(i));
                }
                System.out.println(response.body().string());
            }
        });
```

13.4.2 使用 POST 方法进行服务请求

在使用 POST 方法进行服务请求时，可以将普通字符串、文件、表单等多种格式的请求数据发送到服务器端，并且这些请求数据都是打包到请求包的请求体中的，所以与使用 GET 方法相比，使用 POST 方法更加安全、灵活。

1. 使用 POST 方法提交字符串

使用 POST 方法提交字符串的实现代码如下：

```
    public static final MediaType MEDIA_TYPE_MARKDOWN
                    = MediaType.parse("text/x-markdown; charset=utf-8");
    private final OkHttpClient client = new OkHttpClient();

    public void postString() throws Exception {
        String postBody = ""
                + "Releases\n"
                + "--------\n"
                + "\n"
                + " * _1.0_ May 6, 2013\n"
                + " * _1.1_ June 15, 2013\n"
                + " * _1.2_ August 11, 2013\n";

        Request request = new Request.Builder()
                    .url("https://api.github.com/markdown/raw")
                    .post(RequestBody.create(MEDIA_TYPE_MARKDOWN, postBody))
                    .build();
        Response response = client.newCall(request).execute();
        if (!response.isSuccessful()) throw new IOException("Unexpected code " + response);
        System.out.println(response.body().string());
    }
```

2. 使用 POST 方法提交文件

使用 POST 方法提交文件的实现代码如下：

```java
public static final MediaType MEDIA_TYPE_MARKDOWN
        = MediaType.parse("text/x-markdown; charset=utf-8");
private final OkHttpClient client = new OkHttpClient();

public void postFile() throws Exception {
    File file = new File("README.md");
    Request request = new Request.Builder()
            .url("https://api.github.com/markdown/raw")
            .post(RequestBody.create(MEDIA_TYPE_MARKDOWN, file))
            .build();
    Response response = client.newCall(request).execute();
    if (!response.isSuccessful()) throw new IOException("Unexpected code " + response);
    System.out.println(response.body().string());
}
```

3. 使用 POST 方法提交简单表单

使用 FormEncodingBuilder 构建与 HTML 中的<form>标签具有相同效果的请求体，用于封装请求数据，代码如下：

```java
private final OkHttpClient client = new OkHttpClient();

public void postForm() throws Exception {
    RequestBody formBody = new FormBody.Builder()
            .add("search", "Jurassic Park")
            .build();
    Request request = new Request.Builder()
            .url("https://en.wikipedia.org/w/index.php")
            .post(formBody)
            .build();
    Response response = client.newCall(request).execute();
    if (!response.isSuccessful()) throw new IOException("Unexpected code " + response);
    System.out.println(response.body().string());
}
```

4. 使用 POST 方法提交分块表单

MultipartBody.Builder 可以构建与 HTML 文件上传形式兼容的复杂请求体，多块请求体中的每块请求都是一个 Part。例如，构建一个包含名为 title 的字符串 Part 和一个名为 image 的文件 Part，代码如下：

```java
private static final String IMGUR_CLIENT_ID = "……";
private static final MediaType MEDIA_TYPE_PNG =
```

```
MediaType.parse("image/png");
        private final OkHttpClient client = new OkHttpClient();

        public void postMultiPartForm() throws Exception {
            RequestBody requestBody = new MultipartBody.Builder()
                    .setType(MultipartBody.FORM)
                    .addFormDataPart("title", "Square Logo")
                    .addFormDataPart("image", "logo-square.png",
                        RequestBody.create(MEDIA_TYPE_PNG, new
                        File("website/static/logo-square.png")))
                    .build();
            Request request = new Request.Builder()
                    .header("Authorization", "Client-ID " + IMGUR_CLIENT_ID)
                    .url("https://api.imgur.com/3/image")
                    .post(requestBody)
                    .build();
            Response response = client.newCall(request).execute();
            if (!response.isSuccessful()) throw new IOException("Unexpected code "
+ response);
            System.out.println(response.body().string());
        }
```

13.4.3 构造请求头及读取响应头

典型的 HTTP 头格式为 Map<String, String>，每个字段中都可以有一个值或多个值，也可以没有值。

当构造请求头时，使用 header(name, value)方法可以设置唯一的 name/value 对。如果已经有 value，那么先移除旧的 value，再添加新的 value。使用 addHeader(name, value)方法可以添加多个 value。

当读取响应头时，使用 header(name)方法返回最后出现的 name/value 对，通常这也是唯一的 name/value 对。如果没有 value，那么 header(name)方法会返回 null，如果要读取字段对应的所有 value，则使用 headers(name)方法会返回一个列表。为了获取所有的 Header，Headers 类支持使用 Index 方式访问。具体代码如下：

```
        private final OkHttpClient client = new OkHttpClient();
        public void headerExample() throws Exception {
            Request request = new Request.Builder()
                    .url("https://api.github.com/repos/square/okhttp/issues")
                    .header("User-Agent", "OkHttp Headers.java")
                    .addHeader("Accept", "application/json; q=0.5")
                    .addHeader("Accept", "application/vnd.github.v3+json")
                    .build();
            Response response = client.newCall(request).execute();
```

```
            if (!response.isSuccessful()) throw new IOException("Unexpected code
" + response);
            System.out.println("Server: " + response.header("Server"));
            System.out.println("Date: " + response.header("Date"));
            System.out.println("Vary: " + response.headers("Vary"));
        }
```

13.4.4　配置 OkHttp 超时

通过进行 OkHttp 超时设置，设置当服务器端没有及时响应时，OkHttp 客户端等待服务器端响应的最长等待时长。没有响应的原因可能是客户端连接问题、服务器端可用性问题等。OkHttp 支持连接超时、读取超时和写入超时。创建具有超时属性的 OkHttpClient 对象的代码如下：

```
    private final OkHttpClient client;

    public ConfigureTimeouts() throws Exception {
        client = new OkHttpClient.Builder()
                .connectTimeout(10, TimeUnit.SECONDS)
                .writeTimeout(10, TimeUnit.SECONDS)
                .readTimeout(30, TimeUnit.SECONDS)
                .build();
    }

    public void run() throws Exception {
        Request request = new
Request.Builder().url("http://httpbin.org/delay/2").build();
        Response response = client.newCall(request).execute();
        System.out.println("Response completed: " + response);
    }
```

13.5　OkHttp GET 实现案例

下面举例说明如何使用 OkHttp 的 GET 方法（HTTP 的 GET 请求方法）与服务器端进行 HTTP 通信。本案例应用程序要实现的功能与 13.2 节中案例应用程序实现的功能相同，运行效果也与图 13-3 所示的运行效果相同。

下面构建该案例应用程序。在 Android Studio 中新建一个名为 Ch1303 的 Android 应用程序工程，将 OkHttp 相关包添加到 Ch1303 应用程序工程中。修改布局文件 res/layout/activity_main.xml 中的代码，修改后的代码如下：

```
<?xml version="1.0" encoding="utf-8"?>
<LinearLayout xmlns:android="http://schemas.android.com/apk/res/android"
    android:id="@+id/activity_main"
```

```xml
    android:layout_width="match_parent"
    android:layout_height="match_parent"
    android:orientation="vertical">

    <Button
        android:id="@+id/id_btn_1"
        android:layout_width="match_parent"
        android:layout_height="wrap_content"
        android:text="@string/text_btn_1" />

    <Button
        android:id="@+id/id_btn_2"
        android:layout_width="match_parent"
        android:layout_height="wrap_content"
        android:text="@string/text_btn_2" />

    <Button
        android:id="@+id/id_btn_3"
        android:layout_width="match_parent"
        android:layout_height="wrap_content"
        android:text="@string/text_btn_3" />

    <Button
        android:id="@+id/id_btn_4"
        android:layout_width="match_parent"
        android:layout_height="wrap_content"
        android:text="@string/text_btn_4" />

    <ImageView
        android:id="@+id/id_iv"
        android:layout_width="match_parent"
        android:layout_height="match_parent"
        android:scaleType="fitCenter"
        android:contentDescription="@string/hello_world"
        />

</LinearLayout>
```

这个布局文件很简单，主要包含 4 个 Button 组件和一个 ImageView 组件。

修改 res/values/strings.xml 文件中的代码，在其中定义布局文件中引用的字符串资源，修改后的代码如下：

```xml
<resources>
    <string name="app_name">Ch1303</string>
```

```xml
        <string name="hello_world">Hello world!</string>

        <string name="text_btn_1">显示第一张蝴蝶</string>
        <string name="text_btn_2">显示第二张蝴蝶</string>
        <string name="text_btn_3">显示第一张卡通</string>
        <string name="text_btn_4">显示第二张卡通</string>

</resources>
```

修改 MainActivity.java 文件中的代码,使其显示主界面,监听对按钮的点击事件,然后根据点击的按钮,使用 OkHttp 的 GET 方法从服务器端获取相应的图片,并且将其显示在 ImageView 组件中,修改后的代码如下:

```java
package com.example.ch1303;

import androidx.appcompat.app.AppCompatActivity;

import android.graphics.Bitmap;
import android.graphics.BitmapFactory;
import android.net.ConnectivityManager;
import android.net.NetworkInfo;
import android.os.Bundle;
import android.os.Handler;
import android.view.View;
import android.widget.Button;
import android.widget.ImageView;
import android.widget.Toast;

import java.io.IOException;

import okhttp3.OkHttpClient;
import okhttp3.Request;
import okhttp3.Response;

public class MainActivity extends AppCompatActivity implements View.OnClickListener {
    private ImageView iv;
    private Bitmap bm;
    private Handler handler;

    @Override
    protected void onCreate(Bundle savedInstanceState) {
        super.onCreate(savedInstanceState);
```

```java
        setContentView(R.layout.activity_main);

        bm = null;
        handler = new Handler();

        iv = (ImageView) this.findViewById(R.id.id_iv);

        Button btn_1 = (Button) this.findViewById(R.id.id_btn_1);
        btn_1.setOnClickListener(this);
        Button btn_2 = (Button) this.findViewById(R.id.id_btn_2);
        btn_2.setOnClickListener(this);
        Button btn_3 = (Button) this.findViewById(R.id.id_btn_3);
        btn_3.setOnClickListener(this);
        Button btn_4 = (Button) this.findViewById(R.id.id_btn_4);
        btn_4.setOnClickListener(this);
    }

    private boolean checkNetworkState() {
        ConnectivityManager cm = (ConnectivityManager) this
                .getSystemService(MainActivity.CONNECTIVITY_SERVICE);
        NetworkInfo ni = cm.getActiveNetworkInfo();
        if ((ni == null) || (!ni.isConnected())) {
            return false;
        }

        return true;
    }

    @Override
    public void onClick(View v) {
        if (!checkNetworkState()) {
            Toast.makeText(this, "网络没有打开,请打开网络后再试。",
                    Toast.LENGTH_LONG).show();
            return;
        }

        int id = v.getId();

        switch(id) {
            case R.id.id_btn_1:
                downloadImageAndShow(1, 1);
                break;
            case R.id.id_btn_2:
```

```
                downloadImageAndShow(1, 2);
                break;
            case R.id.id_btn_3:
                downloadImageAndShow(2, 1);
                break;
            case R.id.id_btn_4:
                downloadImageAndShow(2, 2);
                break;
        }
    }

    private void downloadImageAndShow(final int type, final int id) {
        new Thread(new Runnable() {
            @Override
            public void run() {
                OkHttpClient client = new OkHttpClient();

                Request request = new Request.Builder()
                        .url("http://192.168.197.128:8080/an/ImageShower?" +
                                "type=" + type + "&id=" + id).build();

                Response response;
                try {
                    response = client.newCall(request).execute();
                } catch (IOException e) {
                    e.printStackTrace();
                    return;
                }
                if (!response.isSuccessful()) {
                    return;
                }

                byte[] b = null;
                try {
                    b = response.body().bytes();
                } catch (IOException e) {
                    e.printStackTrace();
                    return;
                }
                bm = BitmapFactory.decodeByteArray(b, 0, b.length);
                handler.post(new Runnable(){
                    @Override
                    public void run() {
```

```
                iv.setImageBitmap(bm);
            }
        });
    }
}).start();
```

在对按钮点击事件的响应方法 onClick() 中，应用程序会根据所点击的按钮，调用 downloadImageAndShow() 方法，获取相应的图片并将其显示在 ImageView 组件中。在 downloadImageAndShow() 方法中，使用 Java 的 Thread 线程机制创建获取图片的线程。

在运行 Ch1303 应用程序前，需要在 AndroidManifest.xml 文件中对网络相关的调用进行授权，代码如下：

```
<uses-permission android:name="android.permission.ACCESS_NETWORK_STATE"/>
<uses-permission android:name="android.permission.INTERNET"/>
```

运行 Ch1303 应用程序，即可得到图 13-3 所示的运行效果。

13.6　OkHttp POST 实现案例

与 OkHttp 的 GET 方法将请求参数放置在 URL 地址中不同，OkHttp 的 POST 方法将发送给服务器端的请求参数放置在请求体中。在理论上，使用 POST 方法可以将任意合法格式的数据发送给服务器端。

下面举例说明如何使用 OkHttp 的 POST 方法（HTTP 的 POST 请求方法）与服务器端布局进行 HTTP 通信。本案例应用程序要实现的功能与 13.2 节中案例应用程序要实现的功能相同，运行效果也与图 13-3 所示的运行效果相同。

现在构建该案例应用程序。在 Android Studio 中新建一个名为 Ch1304 的 Android 应用程序工程。修改布局文件 res/layout/activity_main.xml 中的代码，修改后的代码如下：

```
<LinearLayout xmlns:android="http://schemas.android.com/apk/res/android"
    android:layout_width="match_parent"
    android:layout_height="match_parent"
    android:orientation="vertical">

    <Button
        android:id="@+id/id_btn_1"
        android:layout_width="match_parent"
        android:layout_height="wrap_content"
        android:text="@string/text_btn_1" />

    <Button
        android:id="@+id/id_btn_2"
        android:layout_width="match_parent"
        android:layout_height="wrap_content"
```

```xml
            android:text="@string/text_btn_2" />

        <Button
            android:id="@+id/id_btn_3"
            android:layout_width="match_parent"
            android:layout_height="wrap_content"
            android:text="@string/text_btn_3" />

        <Button
            android:id="@+id/id_btn_4"
            android:layout_width="match_parent"
            android:layout_height="wrap_content"
            android:text="@string/text_btn_4" />

        <ImageView
            android:id="@+id/id_iv"
            android:layout_width="match_parent"
            android:layout_height="match_parent"
            android:scaleType="fitCenter"
            android:contentDescription="@string/hello_world"
            />

</LinearLayout>
```

这个布局文件很简单，主要包含 4 个 Button 组件和一个 ImageView 组件。

修改 res/values/strings.xml 文件中的代码，在其中定义布局文件中引用的字符串资源，修改后的代码如下：

```xml
<?xml version="1.0" encoding="utf-8"?>
<resources>

    <string name="app_name">Ch1304</string>
    <string name="hello_world">Hello world!</string>

    <string name="text_btn_1">显示第一张蝴蝶</string>
    <string name="text_btn_2">显示第二张蝴蝶</string>
    <string name="text_btn_3">显示第一张卡通</string>
    <string name="text_btn_4">显示第二张卡通</string>

</resources>
```

修改 MainActivity.java 文件中的代码，使其显示主界面，监听对按钮的点击事件，然后根据点击的按钮，使用 OkHttp 的 POST 方法从服务器端获取相应的图片，并且将其显示在 ImageView 组件中，修改后的代码如下：

```java
package com.example.ch1304;

import androidx.appcompat.app.AppCompatActivity;

import android.graphics.Bitmap;
import android.graphics.BitmapFactory;
import android.net.ConnectivityManager;
import android.net.NetworkInfo;
import android.os.Bundle;
import android.os.Handler;
import android.view.View;
import android.widget.Button;
import android.widget.ImageView;
import android.widget.Toast;

import java.io.IOException;

import okhttp3.Call;
import okhttp3.Callback;
import okhttp3.FormBody;
import okhttp3.OkHttpClient;
import okhttp3.Request;
import okhttp3.RequestBody;
import okhttp3.Response;

public class MainActivity extends AppCompatActivity implements View.OnClickListener {
    private ImageView iv;
    private Bitmap bm;
    private Handler handler;

    @Override
    protected void onCreate(Bundle savedInstanceState) {
        super.onCreate(savedInstanceState);
        setContentView(R.layout.activity_main);

        bm = null;
        handler = new Handler();

        iv = (ImageView) this.findViewById(R.id.id_iv);

        Button btn_1 = (Button) this.findViewById(R.id.id_btn_1);
        btn_1.setOnClickListener(this);
```

```java
        Button btn_2 = (Button) this.findViewById(R.id.id_btn_2);
        btn_2.setOnClickListener(this);
        Button btn_3 = (Button) this.findViewById(R.id.id_btn_3);
        btn_3.setOnClickListener(this);
        Button btn_4 = (Button) this.findViewById(R.id.id_btn_4);
        btn_4.setOnClickListener(this);
    }

    private boolean checkNetworkState() {
        ConnectivityManager cm = (ConnectivityManager) this
                .getSystemService(MainActivity.CONNECTIVITY_SERVICE);
        NetworkInfo ni = cm.getActiveNetworkInfo();
        if ((ni == null) || (!ni.isConnected())) {
            return false;
        }

        return true;
    }

    @Override
    public void onClick(View v) {
        if (!checkNetworkState()) {
            Toast.makeText(this, "网络没有打开，请打开网络后再试。",
                    Toast.LENGTH_LONG).show();
            return;
        }

        int id = v.getId();

        switch(id) {
            case R.id.id_btn_1:
                downloadImageAndShow(1, 1);
                break;
            case R.id.id_btn_2:
                downloadImageAndShow(1, 2);
                break;
            case R.id.id_btn_3:
                downloadImageAndShow(2, 1);
                break;
            case R.id.id_btn_4:
                downloadImageAndShow(2, 2);
                break;
        }
```

```java
    }
    private void downloadImageAndShow(final int type, final int id) {
        new Thread(new Runnable() {
            @Override
            public void run() {
                OkHttpClient client = new OkHttpClient();
                RequestBody formBody = new FormBody.Builder()
                        .add("type", "" + type)
                        .add("id", "" + id)
                        .build();
                Request request = new Request.Builder()
                        .url("http://192.168.197.128:8080/an/ImageShower")
                        .post(formBody)
                        .build();

                Response response;
                client.newCall(request).enqueue(new Callback() {
                    @Override
                    public void onFailure(Call call, IOException e) {
                        e.printStackTrace();
                    }

                    @Override
                    public void onResponse(Call call, Response response)
                            throws IOException {
                        if (!response.isSuccessful()) {
                            return;
                        }

                        byte[] b = null;
                        try {
                            b = response.body().bytes();
                        } catch (IOException e) {
                            e.printStackTrace();
                            return;
                        }
                        bm = BitmapFactory.decodeByteArray(b, 0, b.length);
                        handler.post(new Runnable() {
                            @Override
                            public void run() {
                                iv.setImageBitmap(bm);
                            }
                        });
```

```
                    }
                });
            }
        }).start();
    }
}
```

在对按钮点击事件的响应方法 onClick()中，应用程序会根据所点击的按钮，调用 downloadImageAndShow()方法，获取相应的图片并将其显示在 ImageView 组件中。在 downloadImageAndShow()方法中，使用 Java 的 Thread 线程机制创建获取图片的线程。在这段代码中，我们还使用了 OkHttp 的异步机制获取图片。

在运行 Ch1304 应用程序前，需要在 AndroidManifest.xml 文件中对网络相关的调用进行授权，代码如下：

```
<uses-permission android:name="android.permission.ACCESS_NETWORK_STATE"/>
<uses-permission android:name="android.permission.INTERNET"/>
```

运行 Ch1304 应用程序，即可得到图 13-3 所示的运行效果。

13.7 同步练习二

在任意一个公共网站上，使用 OkHttpClient 的 GET 方法请求一个 HTML 页面，并且将获取的 HTML 页面显示在 WebView 组件中。

13.8 使用 Multipart 传递请求数据到服务器端

使用 Multipart 可以传递包含文件流的请求数据到服务器端。例如，在一个注册应用程序中，如果需要传递包括头像在内的注册信息到服务器端，则需要使用 Multipart 数据体格式的请求数据。

下面举例说明如何使用 Multipart 格式从 Android 应用程序向服务器端传递 Multipart 数据体格式的请求数据。该案例应用程序的运行效果与图 13-3 所示的运行效果相同，只是实现的方式不同。为了演示 Multipart 的数据请求，我们在请求数据包中附加了一个简单的图片资源文件，服务器端在收到这个图片资源文件后，将其存储于其工作目录下的 images 目录下，图片资源文件名为 photo.png。因此，需要在服务器端创建一个支持 Multipart 数据体格式请求数据的 Servlet，并且将该 Servlet 命名为 ImageShowerMultipart，该 Servlet 的代码如下：

```
package com.ttt.servlet;

import java.io.ByteArrayOutputStream;
import java.io.FileInputStream;
import java.io.FileOutputStream;
import java.io.IOException;
import java.io.InputStream;
```

```java
import javax.servlet.ServletException;
import javax.servlet.ServletOutputStream;
import javax.servlet.annotation.MultipartConfig;
import javax.servlet.annotation.WebServlet;
import javax.servlet.http.HttpServlet;
import javax.servlet.http.HttpServletRequest;
import javax.servlet.http.HttpServletResponse;
import javax.servlet.http.Part;

import org.apache.commons.fileupload.servlet.ServletFileUpload;

@WebServlet("/ImageShowerMultipart")
@MultipartConfig
public class ImageShowerMultipart extends HttpServlet {
    private static final long serialVersionUID = 1L;

    protected void doGet(HttpServletRequest request, HttpServletResponse response)
            throws ServletException, IOException {
        request.setCharacterEncoding("utf-8");

        System.out.println("GET");

        String type = request.getParameter("type");
        if ((type == null) || (type.equalsIgnoreCase(""))) {
            type = "1";
        }
        String id = request.getParameter("id");
        if ((id == null) || (id.equalsIgnoreCase(""))) {
            id = "1";
        }

        FileInputStream fis =
                new FileInputStream(this.getServletContext().getRealPath("") +
                "images/png" + type + id + ".png");
        byte[] b=new byte[fis.available()];
        fis.read(b);
        fis.close();

        response.setContentType("image/png");
        ServletOutputStream op = response.getOutputStream();
        op.write(b);
        op.close();
    }
```

```java
    protected void doPost(HttpServletRequest request, HttpServletResponse response)
            throws ServletException, IOException {
        request.setCharacterEncoding("utf-8");

        System.out.println("POST");

        if (ServletFileUpload.isMultipartContent(request)) {
            Part part = request.getPart("image");
            InputStream is = part.getInputStream();

            ByteArrayOutputStream baos = new ByteArrayOutputStream();

            byte[] b = new byte[1024];
            while(is.read(b)>0) {
                baos.write(b);
            }

            b = baos.toByteArray();

            FileOutputStream fos = new
                    FileOutputStream(this.getServletContext().getRealPath("") +
                    "images/photo.png");
            fos.write(b);
            fos.close();
        }

        doGet(request, response);
    }
}
```

在该 Servlet 中，为了处理 Multipart 数据体格式的请求数据，我们为该 Servlet 添加了以下标注。在 doPost()方法中，判断是否为 Multipart 数据体格式的请求数据，若是，则从请求数据包中获取数据流并将其存储于该 Web 工程的 images/photo.png 文件中。

```
@MultipartConfig
```

现在编写客户端程序代码。在 Android Studio 中新建一个名为 Ch1305 的 Android 应用程序工程，修改布局文件 res/layout/activity_main.xml 中的代码，修改后的代码如下：

```xml
<?xml version="1.0" encoding="utf-8"?>
<LinearLayout xmlns:android="http://schemas.android.com/apk/res/android"
    android:layout_width="match_parent"
    android:layout_height="match_parent"
    android:orientation="vertical">
```

```xml
<Button
    android:id="@+id/id_btn_1"
    android:layout_width="match_parent"
    android:layout_height="wrap_content"
    android:text="@string/text_btn_1" />

<Button
    android:id="@+id/id_btn_2"
    android:layout_width="match_parent"
    android:layout_height="wrap_content"
    android:text="@string/text_btn_2" />

<Button
    android:id="@+id/id_btn_3"
    android:layout_width="match_parent"
    android:layout_height="wrap_content"
    android:text="@string/text_btn_3" />

<Button
    android:id="@+id/id_btn_4"
    android:layout_width="match_parent"
    android:layout_height="wrap_content"
    android:text="@string/text_btn_4" />

<ImageView
    android:id="@+id/id_iv"
    android:layout_width="match_parent"
    android:layout_height="match_parent"
    android:scaleType="fitCenter"
    android:contentDescription="@string/hello_world"
    />

</LinearLayout>
```

修改 res/values/strings.xml 文件中的代码，在其中定义布局文件中引用的字符串资源，修改后的代码如下：

```xml
<resources>
    <string name="app_name">Ch1305</string>

    <string name="hello_world">Hello world!</string>

    <string name="text_btn_1">显示第一张蝴蝶</string>
```

```xml
    <string name="text_btn_2">显示第二张蝴蝶</string>
    <string name="text_btn_3">显示第一张卡通</string>
    <string name="text_btn_4">显示第二张卡通</string>

</resources>
```

修改 MainActivity.java 文件中的代码，修改后的代码如下：

```java
package com.example.ch1305;

import androidx.appcompat.app.AppCompatActivity;

import android.graphics.Bitmap;
import android.graphics.BitmapFactory;
import android.net.ConnectivityManager;
import android.net.NetworkInfo;
import android.os.Bundle;
import android.os.Environment;
import android.os.Handler;
import android.view.View;
import android.widget.Button;
import android.widget.ImageView;
import android.widget.Toast;

import java.io.File;
import java.io.IOException;

import okhttp3.Call;
import okhttp3.Callback;
import okhttp3.MediaType;
import okhttp3.MultipartBody;
import okhttp3.OkHttpClient;
import okhttp3.Request;
import okhttp3.RequestBody;
import okhttp3.Response;

public class MainActivity extends AppCompatActivity implements View.OnClickListener {
    private ImageView iv;
    private Bitmap bm;
    private Handler handler;

    @Override
```

```java
protected void onCreate(Bundle savedInstanceState) {
    super.onCreate(savedInstanceState);
    setContentView(R.layout.activity_main);

    bm = null;
    handler = new Handler();

    iv = (ImageView) this.findViewById(R.id.id_iv);

    Button btn_1 = (Button) this.findViewById(R.id.id_btn_1);
    btn_1.setOnClickListener(this);
    Button btn_2 = (Button) this.findViewById(R.id.id_btn_2);
    btn_2.setOnClickListener(this);
    Button btn_3 = (Button) this.findViewById(R.id.id_btn_3);
    btn_3.setOnClickListener(this);
    Button btn_4 = (Button) this.findViewById(R.id.id_btn_4);
    btn_4.setOnClickListener(this);
}

private boolean checkNetworkState() {
    ConnectivityManager cm = (ConnectivityManager) this
            .getSystemService(MainActivity.CONNECTIVITY_SERVICE);
    NetworkInfo ni = cm.getActiveNetworkInfo();
    if ((ni == null) || (!ni.isConnected())) {
        return false;
    }

    return true;
}

@Override
public void onClick(View v) {
    if (!checkNetworkState()) {
        Toast.makeText(this, "网络没有打开，请打开网络后再试。",
                Toast.LENGTH_LONG).show();
        return;
    }

    int id = v.getId();

    switch(id) {
```

```java
            case R.id.id_btn_1:
                downloadImageAndShow(1, 1);
                break;
            case R.id.id_btn_2:
                downloadImageAndShow(1, 2);
                break;
            case R.id.id_btn_3:
                downloadImageAndShow(2, 1);
                break;
            case R.id.id_btn_4:
                downloadImageAndShow(2, 2);
                break;
        }
    }

    private void downloadImageAndShow(final int type, final int id) {
        new Thread(new Runnable() {
            @Override
            public void run() {
                OkHttpClient client = new OkHttpClient();

                MediaType MEDIA_TYPE_PNG = MediaType.parse("image/png");
                RequestBody requestBody = new MultipartBody.Builder()
                        .setType(MultipartBody.FORM)
                        .addFormDataPart("type", ""+type)
                        .addFormDataPart("id", ""+id)
                        .addFormDataPart("image", "photo.png",
                                RequestBody.create(MEDIA_TYPE_PNG,
                                        new
                                    File(Environment.getExternalStorageDirectory() +
                                            "/photo.png"))).build();

                Request request = new Request.Builder()
                        .url("http://192.168.197.128:8080/an/ImageShowerMultipart")
                                .post(requestBody)
                                .build();

                client.newCall(request).enqueue(new Callback() {
                    @Override
                    public void onFailure(Call call, IOException e) {
                        e.printStackTrace();
```

```java
        }

        @Override
        public void onResponse(Call call, Response response)
            throws IOException {
            if (!response.isSuccessful()) {
                return;
            }

            byte[] b = null;
            try {
                b = response.body().bytes();
            } catch (IOException e) {
                e.printStackTrace();
                return;
            }
            bm = BitmapFactory.decodeByteArray(b, 0, b.length);
            handler.post(new Runnable() {
                @Override
                public void run() {
                    iv.setImageBitmap(bm);
                }
            });
        }
    });
  }
}).start();
```

与 Ch1304 应用程序不同的是，在 Ch1305 应用程序中，我们采用 Multipart 封装请求数据，代码如下：

```java
RequestBody requestBody = new MultipartBody.Builder()
        .setType(MultipartBody.FORM)
        .addFormDataPart("type", ""+type)
        .addFormDataPart("id", ""+id)
        .addFormDataPart("image", "photo.png",
            RequestBody.create(MEDIA_TYPE_PNG,
                new File(Environment.getExternalStorageDirectory() +
                    "/photo.png")))
        .build();
```

Multipart 中包含 3 个请求参数，包括名字为 type 的字符串、名字为 id 的字符串和一个文件。

要运行 Ch1305 应用程序，还需要在 AndroidManifest.xml 文件中添加以下代码，用于对网络进行相应的授权申请。

```
<uses-permission android:name="android.permission.ACCESS_NETWORK_STATE" />
<uses-permission android:name="android.permission.INTERNET" />
<uses-permission android:name="android.permission.READ_EXTERNAL_STORAGE" />
```

运行 Ch1305 应用程序，即可得到图 13-3 所示的运行效果。观察服务器端，在 Web 的工作目录下会生成一个用于存储来自客户端的图片资源文件的目录。

13.9 同步练习三

使用 OkHttp 编写一个进行网络信息注册的应用程序，包括客户端程序和服务器端程序，注册信息包括姓名、出生日期、密码、电话号码和头像。由于请求数据中包含头像，因此只能使用 POST 方法发送请求数据。

13.10 使用 JSON 格式的数据与服务器端通信

13.10.1 JSON 基础

JSON 是一种数据交换的标准格式。本节主要介绍 JSON 的概念和应用。

JSON 是一种轻量级的数据交换格式，它使用 name/value 对表示数据。JSON 支持两种结构：以"{}"（大括号）表示的对象数据和以"[]"（中括号）表示的数组数据。JSON 表示数据的能力在于，它可以使用两种基础格式的组合表示任意复杂的数据。例如，使用 JSON 格式的数据（简称 JSON 数据）表示一个人的基本信息，代码如下：

```
{
    "name": "Geoge Bush",
    "age": 20,
    "memo": "乔治毕业于哈佛大学，获得计算机科学博士学位……",
    "phone": "13800138000"
}
```

上述 JSON 数据表示的信息非常清楚：一个人，他的名字为 Geoge Bush，他的年龄为 20 岁，他的简介为"乔治毕业于哈佛大学，获得计算机科学博士学位……"，他的电话号码为 13800138000。要表示多个人的基本信息，可以使用中括号将多个人的基本信息括起来，示例代码如下：

```
[
    {
        "name": "Geoge Bush",
        "age": 20,
        "memo": "乔治毕业于哈佛大学，获得计算机科学博士学位……",
```

```
        "phone": "13800138000"
    },
    {
        "name": "Bill Gates",
        "age": 23,
        "memo": "比尔……",
        "phone": "13800138001"
    }
    …
]
```

JSON 数据的格式不同，访问 JSON 数据的方式也有所不同。如果一个 JSON 数据是对象数据，则使用"变量名.成员名"的方式访问该 JSON 数据。例如，一个 JSON 对象的名称为 var1，可以使用 var1.name 的方式访问 name 属性的值。如果一个 JSON 数据为数组数据，则使用"变量名[下标].成员名"的方式访问该 JSON 数据。例如，JSON 数组的名称为 var2，可以使用 var2[2].name 的方式访问第 2 个人的 name 属性。

JSON 数据属性的数据类型可以是计算机支持的任意数据类型，包括数字、字符串、逻辑值（true 或 false）、数组、对象、null。下面来看一个复杂的 JSON 数据，代码如下：

```
{
    "name": "Bill Gates",
    "workday": ["Monday", "Tuesday", "Friday"],
    "salary": 8700.5,
    "birth": "1980-10-10",
    "memo": "Something…"
    "alive": true
}
```

13.10.2 在 JavaScript 中使用 JSON 数据

JSON 是 JavaScript 支持的原生数据格式，因此在 JavaScript 中使用 JSON 数据非常简单。例如，在 JavaScript 中，可以直接设置变量的值为一个 JSON 数据，代码如下：

```
var bill =
    {
        "name": "Bill Gates",
        "workday": ["Monday", "Tuesday", "Friday"],
        "salary": 8700.5,
        "birth": "1980-10-10",
        "memo": "Something…"
        "alive": true
    }
```

13.10.3 在 Java 中使用 JSON 数据

Java 不能直接支持 JSON 数据。在 Java 中，一个 JSON 数据被看作一个字符串，称为 JSON 串。通过使用第三方提供的 Jar 包，可以将 JSON 串转换为一个 Java 对象，也可以将一个 Java

对象转换为一个 JSON 串。在这些第三方包中，目前较好用且使用较广泛的是 Google 提供的 Gson 包。可以在 Google 的开发者网站上下载 Gson 包。我们使用 Gson 包的 2.3.1 版本，即 gson-2.3.1.jar 包。在下载完毕后，将 gson-2.3.1.jar 包复制到工程的 libs 目录下，然后选中该包并右击，在弹出的快捷菜单中选择 Add as Library 命令，就可以使用 Gson 包了。

现在我们简单地演示一下如何使用 Gson 包进行 JSON 串与 Java 对象之间的转换。首先，自定义一个 Java 类，如定义一个 Person 类，代码如下：

```java
public class Person {
    public String name;
    public int age;
    public String memo;
    public String phone;
}
```

基于 Person 类，将一个 JSON 串转换为 Person 对象，以及将一个 Person 对象转换为 JSON 串，代码如下：

```java
Gson gson = new GsonBuilder().setDateFormat("yyyy-MM-dd").create();

String bill_json =
"{" +
    "\"name\":" + "\"Bill Gates\"" + "," +
    "\"age\":" + "20" + "," +
    "\"memo\":" + "\"比尔毕业于哈佛大学……\"" + "," +
    "\"phone\":" + "\"13800138000\"" +
"}";
Person bill = gson.fromJson(bill_json, Person.class);
System.out.println(bill.name + "\n" + bill.age + "\n" + bill.phone + "\n" + bill.memo);

Person geoge = new Person();
geoge.name = "Geoge";
geoge.age = 23;
geoge.phone = "13800138001";
geoge.memo = "Something to say…";
String geoge_json = gson.toJson(geoge);
System.out.println(geoge_json);
```

在上述代码中，首先获取一个 Gson 对象，然后将一个 JSON 串转换为 Person 对象，最后将一个 Person 对象转换为 JSON 串。运行上述代码片段，运行效果如图 13-7 所示。

```
Bill Gates
20
13800138000
比尔毕业于哈佛大学……
{"memo":"Something to say_","name":"Geoge","phone":"13800138001","age":23}
```

图 13-7　使用 Gson 包进行 Java 对象和 JSON 串之间转换的示例运行效果

13.10.4　使用 POST 方法及 JSON 数据格式发送请求

下面举例说明如何使用 HTTP 的 POST 方法和 JSON 数据格式与服务器端进行 HTTP 通信。本案例应用程序要实现的功能与 13.2 节中案例应用程序要实现的功能相同，运行效果也与图 13-3 所示的运行效果相同。

为了处理 JSON 格式的数据请求，在服务器端创建一个名为 ImageShowerJSON 的 Servlet，该 Servlet 中的代码如下：

```java
package com.ttt.servlet;

import java.io.FileInputStream;
import java.io.IOException;

import javax.servlet.ServletException;
import javax.servlet.ServletOutputStream;
import javax.servlet.annotation.WebServlet;
import javax.servlet.http.HttpServlet;
import javax.servlet.http.HttpServletRequest;
import javax.servlet.http.HttpServletResponse;

import com.google.gson.Gson;
import com.google.gson.GsonBuilder;

@WebServlet("/ImageShowerJSON")
public class ImageShowerJSON extends HttpServlet {
    private static final long serialVersionUID = 1L;

    protected void doGet(HttpServletRequest request, HttpServletResponse response)
                        throws ServletException, IOException {
        request.setCharacterEncoding("utf-8");
        Gson gson = new GsonBuilder().setDateFormat("yyyy-MM-dd").create();
        String data = request.getParameter("data");
        TypeAndId ti = gson.fromJson(data, TypeAndId.class);

        FileInputStream fis = new FileInputStream(
            this.getServletContext().getRealPath("") +
                        "images/png" + ti.type + ti.id + ".png");
        byte[] b=new byte[fis.available()];
        fis.read(b);
        fis.close();

        response.setContentType("image/png");
        ServletOutputStream op = response.getOutputStream();
```

```java
            op.write(b);
            op.close();
        }

        protected void doPost(HttpServletRequest request, HttpServletResponse response)
                                        throws ServletException, IOException {
            doGet(request, response);
        }

        private class TypeAndId {
            public int type;
            public int id;
        }
    }
```

在 Android Studio 中新建一个名为 Ch1306 的 Android 应用程序工程，并且将 gson-2.8.0.jar 包复制到该工程的 libs 目录下。修改布局文件 res/layout/activity_main.xml 中的代码，修改后的代码如下：

```xml
<?xml version="1.0" encoding="utf-8"?>
<LinearLayout xmlns:android="http://schemas.android.com/apk/res/android"
    android:layout_width="match_parent"
    android:layout_height="match_parent"
    android:orientation="vertical">

    <Button
        android:id="@+id/id_btn_1"
        android:layout_width="match_parent"
        android:layout_height="wrap_content"
        android:text="@string/text_btn_1" />

    <Button
        android:id="@+id/id_btn_2"
        android:layout_width="match_parent"
        android:layout_height="wrap_content"
        android:text="@string/text_btn_2" />

    <Button
        android:id="@+id/id_btn_3"
        android:layout_width="match_parent"
        android:layout_height="wrap_content"
        android:text="@string/text_btn_3" />

    <Button
```

```xml
        android:id="@+id/id_btn_4"
        android:layout_width="match_parent"
        android:layout_height="wrap_content"
        android:text="@string/text_btn_4" />

    <ImageView
        android:id="@+id/id_iv"
        android:layout_width="match_parent"
        android:layout_height="match_parent"
        android:scaleType="fitCenter"
        android:contentDescription="@string/hello_world"
        />

</LinearLayout>
```

修改 res/values/strings.xml 文件中的代码，修改后的代码如下：

```xml
<resources>
    <string name="app_name">Ch1306</string>

    <string name="hello_world">Hello world!</string>

    <string name="text_btn_1">显示第一张蝴蝶</string>
    <string name="text_btn_2">显示第二张蝴蝶</string>
    <string name="text_btn_3">显示第一张卡通</string>
    <string name="text_btn_4">显示第二张卡通</string>

</resources>
```

修改 MainActivity.java 文件中的代码，修改后的代码如下：

```java
package com.example.ch1306;

import androidx.appcompat.app.AppCompatActivity;

import android.graphics.Bitmap;
import android.graphics.BitmapFactory;
import android.net.ConnectivityManager;
import android.net.NetworkInfo;
import android.os.Bundle;
import android.os.Handler;
import android.view.View;
import android.widget.Button;
import android.widget.ImageView;
import android.widget.Toast;
```

```java
import com.google.gson.Gson;
import com.google.gson.GsonBuilder;

import java.io.IOException;

import okhttp3.Call;
import okhttp3.Callback;
import okhttp3.FormBody;
import okhttp3.OkHttpClient;
import okhttp3.Request;
import okhttp3.RequestBody;
import okhttp3.Response;

public class MainActivity extends AppCompatActivity implements View.OnClickListener {
    private ImageView iv;
    private Bitmap bm;
    private Handler handler;

    @Override
    protected void onCreate(Bundle savedInstanceState) {
        super.onCreate(savedInstanceState);
        setContentView(R.layout.activity_main);

        bm = null;
        handler = new Handler();

        iv = (ImageView) this.findViewById(R.id.id_iv);

        Button btn_1 = (Button) this.findViewById(R.id.id_btn_1);
        btn_1.setOnClickListener(this);
        Button btn_2 = (Button) this.findViewById(R.id.id_btn_2);
        btn_2.setOnClickListener(this);
        Button btn_3 = (Button) this.findViewById(R.id.id_btn_3);
        btn_3.setOnClickListener(this);
        Button btn_4 = (Button) this.findViewById(R.id.id_btn_4);
        btn_4.setOnClickListener(this);
    }

    private boolean checkNetworkState() {
        ConnectivityManager cm = (ConnectivityManager) this
                .getSystemService(MainActivity.CONNECTIVITY_SERVICE);
        NetworkInfo ni = cm.getActiveNetworkInfo();
```

```java
        if ((ni == null) || (!ni.isConnected())) {
            return false;
        }

        return true;
    }


    @Override
    public void onClick(View v) {
        if (!checkNetworkState()) {
            Toast.makeText(this, "网络没有打开,请打开网络后再试。",
                    Toast.LENGTH_LONG).show();
            return;
        }

        int id = v.getId();

        switch(id) {
            case R.id.id_btn_1:
                downloadImageAndShow(1, 1);
                break;
            case R.id.id_btn_2:
                downloadImageAndShow(1, 2);
                break;
            case R.id.id_btn_3:
                downloadImageAndShow(2, 1);
                break;
            case R.id.id_btn_4:
                downloadImageAndShow(2, 2);
                break;
        }
    }

    private void downloadImageAndShow(final int type, final int id) {
        new Thread(new Runnable() {
            @Override
            public void run() {
                OkHttpClient client = new OkHttpClient();

                Gson gson = new GsonBuilder().setDateFormat("yyyy-MM-dd").create();
                TypeAndId ti = new TypeAndId();
                ti.type = type;
                ti.id = id;
```

```java
            String json = gson.toJson(ti);

            RequestBody formBody = new FormBody.Builder()
                    .add("data", json)
                    .build();
            Request request = new Request.Builder()
                    .url("http://192.168.197.128:8080/an/ImageShowerJSON")
                    .post(formBody)
                    .build();

            client.newCall(request).enqueue(new Callback() {
                @Override
                public void onFailure(Call call, IOException e) {
                    e.printStackTrace();
                }

                @Override
                public void onResponse(Call call, Response response)
                        throws IOException {
                    if (!response.isSuccessful()) {
                        return;
                    }

                    byte[] b = null;
                    try {
                        b = response.body().bytes();
                    } catch (IOException e) {
                        e.printStackTrace();
                        return;
                    }
                    bm = BitmapFactory.decodeByteArray(b, 0, b.length);
                    handler.post(new Runnable() {
                        @Override
                        public void run() {
                            iv.setImageBitmap(bm);
                        }
                    });
                }
            });
        }
    }).start();
}
```

```
    @SuppressWarnings("unused")
    private class TypeAndId {
        public int type;
        public int id;
    }
}
```

在 MainActivity 类中定义一个 TypeAndId 类，用于组装发送到服务器端获取指定图片资源的文件的参数。在 downloadImageAndShow()方法中，使用以下代码，首先获取一个 Gson 对象，然后将图片资源文件的 type 和 id 封装到 TypeAndId 对象中，再使用 Gson 包将 TypeAndID 对象转换为 JSON 串，最后将 JSON 串作为 data 的参数发送给服务器端。

```
Gson gson = new GsonBuilder().setDateFormat("yyyy-MM-dd").create();
TypeAndId ti = new TypeAndId();
ti.type = type;
ti.id = id;
String json = gson.toJson(ti);
RequestBody formBody = new FormBody.Builder()
        .add("data", json)
        .build();
```

在运行 Ch1306 应用程序前，需要在 AndroidManifest.xml 文件中申请相应的权限，代码如下：

```
<uses-permission android:name="android.permission.ACCESS_NETWORK_STATE"/>
<uses-permission android:name="android.permission.INTERNET"/>
```

运行 Ch1306 应用程序，即可得到图 13-3 所示的运行效果。

第 14 章

Android 和 HTML5 的混合开发

近几年,基于 HTML5 的应用程序开发越来越受青睐。在 Android 平台上,除了可以使用原生 Android 进行应用程序开发(Android 原生开发),还可以将 Android 和 HTML5 相结合,进行混合式应用程序开发(Android 和 HTML5 的混合开发)。与 Android 原生开发相比,Android 和 HTML5 混合开发的优点在于,对 Android 应用程序做较小的修改,即可使其既可以运行于 Android 平台上,又可以运行于普通的浏览器上。本章主要介绍 Android 和 HTML5 的混合开发。

14.1 Android 和 HTML5 的混合开发基础

在 Android 平台上,有一个用于显示网页的基本组件 WebView。使用 WebView 组件,应用程序既可以显示网络 HTML 页面,又可以显示内置在 Android App 中的 HTML 页面。下面举例说明 WebView 组件的使用方法。

在 Android Studio 中新建一个名为 Ch1401 的 Android 应用程序工程,修改布局文件 activity_main.xml 中的代码,修改后的代码如下:

```xml
<?xml version="1.0" encoding="utf-8"?>
<LinearLayout
    xmlns:android="http://schemas.android.com/apk/res/android"
    android:layout_width="match_parent"
    android:layout_height="match_parent"
    android:orientation="vertical">

    <WebView
        android:id="@+id/id_web_view"
        android:layout_width="match_parent"
        android:layout_height="match_parent"
        />

</LinearLayout>
```

这个布局文件很简单,只包含一个 WebView 组件。

为了在 WebView 组件中显示一个 HTML 页面,修改 MainActivity.java 文件中的代码,修改后的代码如下:

```java
package com.example.ch1401;

import androidx.appcompat.app.AppCompatActivity;

import android.os.Bundle;
import android.webkit.WebView;
import android.webkit.WebViewClient;

public class MainActivity extends AppCompatActivity {
    private WebView mWebView;

    @Override
    protected void onCreate(Bundle savedInstanceState) {
        super.onCreate(savedInstanceState);
        setContentView(R.layout.activity_main);

        mWebView = findViewById(R.id.id_web_view);
        mWebView.setWebViewClient(new WebViewClient());
        mWebView.loadUrl("https://www.baidu.com");
    }
}
```

在 MainActivity 类中，首先使用 WebView 组件打开 HTML 页面，而不是使用 Android 手机中的默认浏览器打开 HTML 页面，代码如下：

```java
        mWebView.setWebViewClient(new WebViewClient());
```

然后打开百度页面，代码如下：

```java
        mWebView.loadUrl("https://www.baidu.com");
```

由于该案例应用程序需要访问网络，因此需要在 AndroidManifest.xml 文件中为网络访问授权。修改后的 AndroidManifest.xml 文件中的代码如下：

```xml
<?xml version="1.0" encoding="utf-8"?>
<manifest xmlns:android="http://schemas.android.com/apk/res/android"
    package="com.example.ch1401">

    <uses-permission android:name="android.permission.INTERNET" />

    <application
        android:allowBackup="true"
        android:icon="@mipmap/ic_launcher"
        android:label="@string/app_name"
        android:roundIcon="@mipmap/ic_launcher_round"
        android:supportsRtl="true"
        android:theme="@style/Theme.Ch1401">
```

```xml
        <activity
            android:name=".MainActivity"
            android:exported="true">
            <intent-filter>
                <action android:name="android.intent.action.MAIN" />

                <category android:name="android.intent.category.LAUNCHER" />
            </intent-filter>
        </activity>
    </application>

</manifest>
```

运行 Ch1401 应用程序，即可打开百度页面。

14.2 使用 WebView 组件显示本地页面

使用 WebView 组件不仅可以显示网络 HTML 页面，还可以显示本地 HTML 页面，即内置在 Android App 中的 HTML 页面。为了能够显示内置在 Android App 中的 HTML 页面，Android 要求所有的 HTML 页面必须存储于工程的 src/main/assets 目录下。在默认情况下，src/main/assets 目录是不存在的。因此，需要自行创建 assets 目录。在 Android Studio 中切换到 Project 视图，如图 14-1 所示。

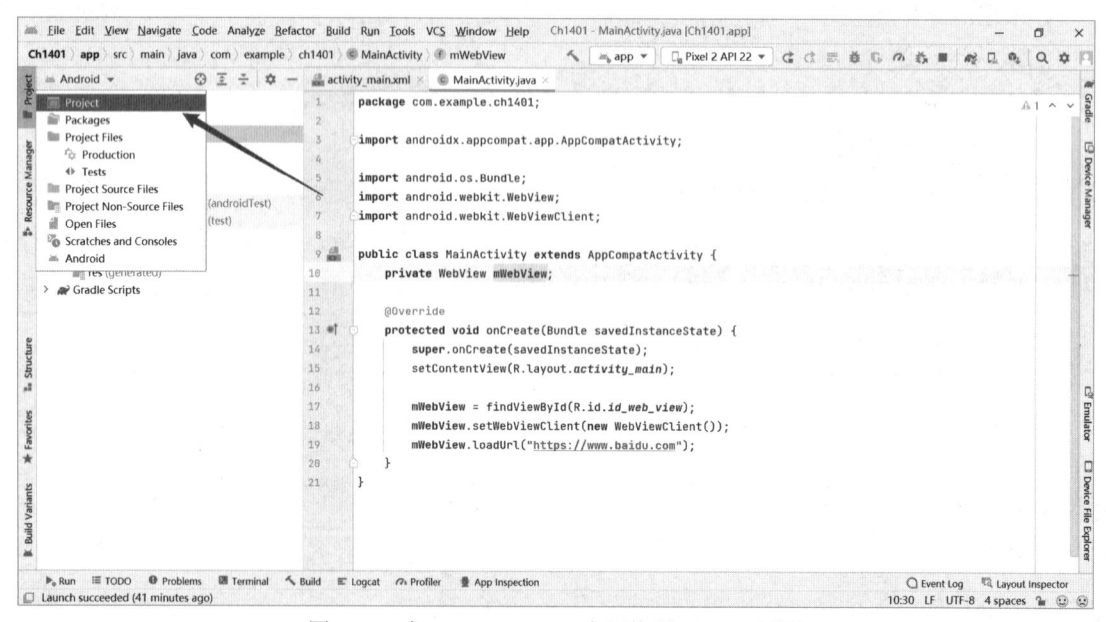

图 14-1　在 Android Studio 中切换到 Project 视图

在 Project 视图中的 src→main 节点下创建 assets 目录，如图 14-2 所示。

在 assets 目录创建完成后，即可在该目录下创建各种所需的子目录。例如，创建 html 子目录，用于存储 HTML 页面文件；创建 images 子目录，用于存储 HTML 页面文件使用的图片。

第 14 章 Android 和 HTML5 的混合开发

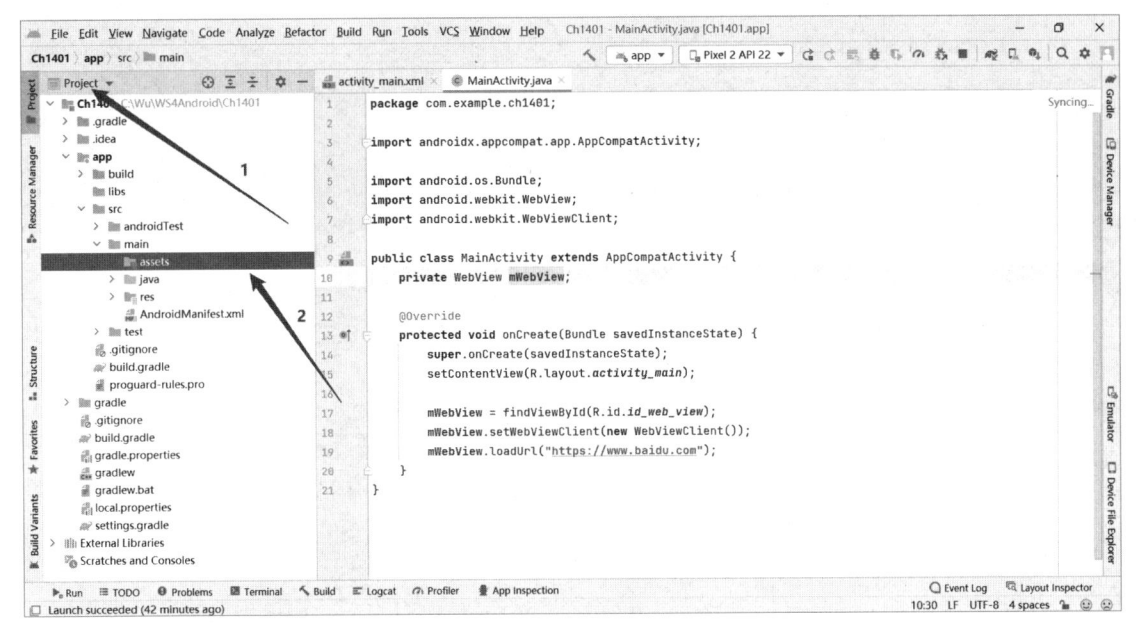

图 14-2 创建 assets 目录

在 src/main/assets 目录下创建 html 子目录、images 子目录和 js 子目录，然后在 html 子目录下创建一个 HTML 页面文件 first.html，并且在 images 子目录下存储 first.html 文件中用到的图片；在 js 子目录下存储 jQuery 文件。构建完成的 assets 目录结构如图 14-3 所示。

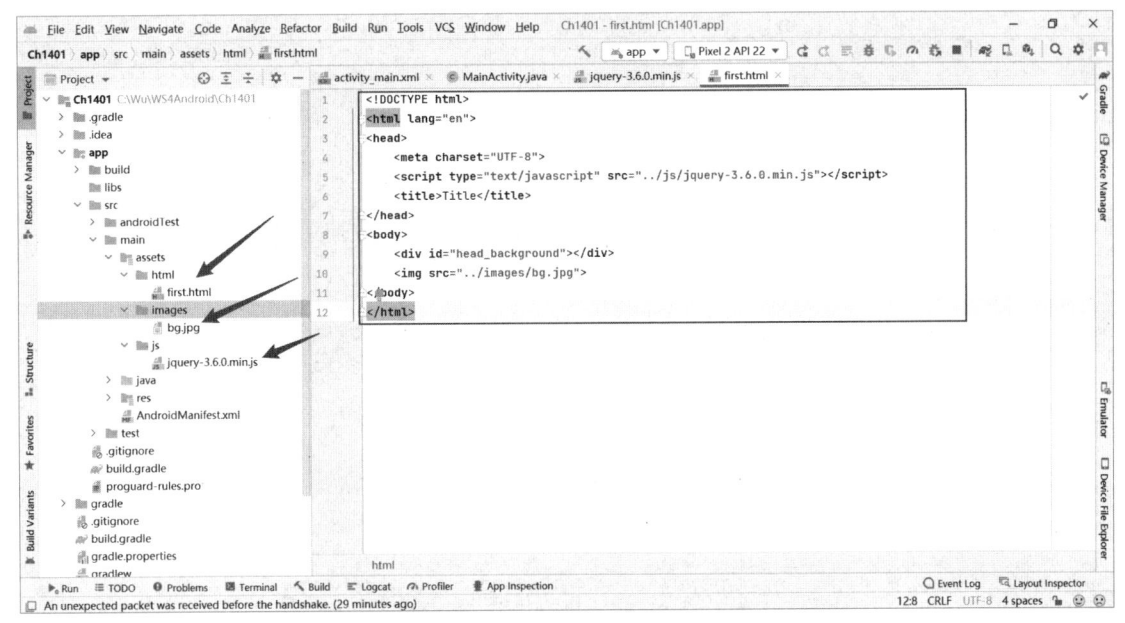

图 14-3 构建完成的 assets 目录结构

first.html 文件中的代码如下：

```
<!DOCTYPE html>
<html lang="en">
<head>
    <meta charset="UTF-8">
```

```html
    <script type="text/javascript" src="../js/jquery-3.6.0.min.js"></script>
    <title>Title</title>
</head>
<body>
    <div id="head_background"></div>
    <img src="../images/bg.jpg">
</body>
</html>
```

修改 MainActivity.java 文件中的代码，使其可以显示 first.html 文件中的内容，修改后的代码如下：

```java
package com.example.ch1401;

import androidx.appcompat.app.AppCompatActivity;

import android.os.Bundle;
import android.webkit.WebView;
import android.webkit.WebViewClient;

public class MainActivity extends AppCompatActivity {
    private WebView mWebView;

    @Override
    protected void onCreate(Bundle savedInstanceState) {
        super.onCreate(savedInstanceState);
        setContentView(R.layout.activity_main);

        mWebView = findViewById(R.id.id_web_view);
        mWebView.setWebViewClient(new WebViewClient());
        //mWebView.loadUrl("https://www.baidu.com");
        mWebView.loadUrl("file:///android_asset/html/first.html");
    }
}
```

通过以下代码，即可在 WebView 组件中显示指定的 HTML 页面。

```java
mWebView.loadUrl("file:///android_asset/html/first.html");
```

14.3 Android 与 HTML5 页面之间的信息交互

在 Android 和 HTML5 的混合开发中，不可避免地会要求 Android 的原生 Java 代码与 HTML 页面进行信息交互，这种信息交互包括两方面：其一，从 Android 的原生 Java 代码向 HTML 页面传送信息，如将 Java 处理的信息显示在 HTML 页面上；其二，从 HTML 页面向 Android 的原生 Java 代码传送数据，这些数据需要在 Android 的原生 Java 代码中进行进一步的处理。

第 14 章 Android 和 HTML5 的混合开发

下面举例说明如何在 Android 的原生 Java 代码与 HTML 页面之间进行信息交互。

在该案例应用程序中，首先启动一个 Activity，该 Activity 中包含一个输入框和一个按钮，用户在输入框中输入信息后，点击按钮，即可启动一个新的 Activity，这个新的 Activity 通过 WebView 组件打开一个 HTML 页面，在该 HTML 页面中显示一张图片和从 Android 原生 Java 代码传递来的信息，点击 HTML 页面中的信息，可以再启动一个新的 Activity，在这个新的 Activity 中显示 HTML 页面中的信息。

在 Android Studio 中新建一个名为 Ch1402 的应用程序工程，按照 14.2 节中的步骤创建 assets 目录。创建完成的 Ch1402 应用程序工程目录如图 14-4 所示。

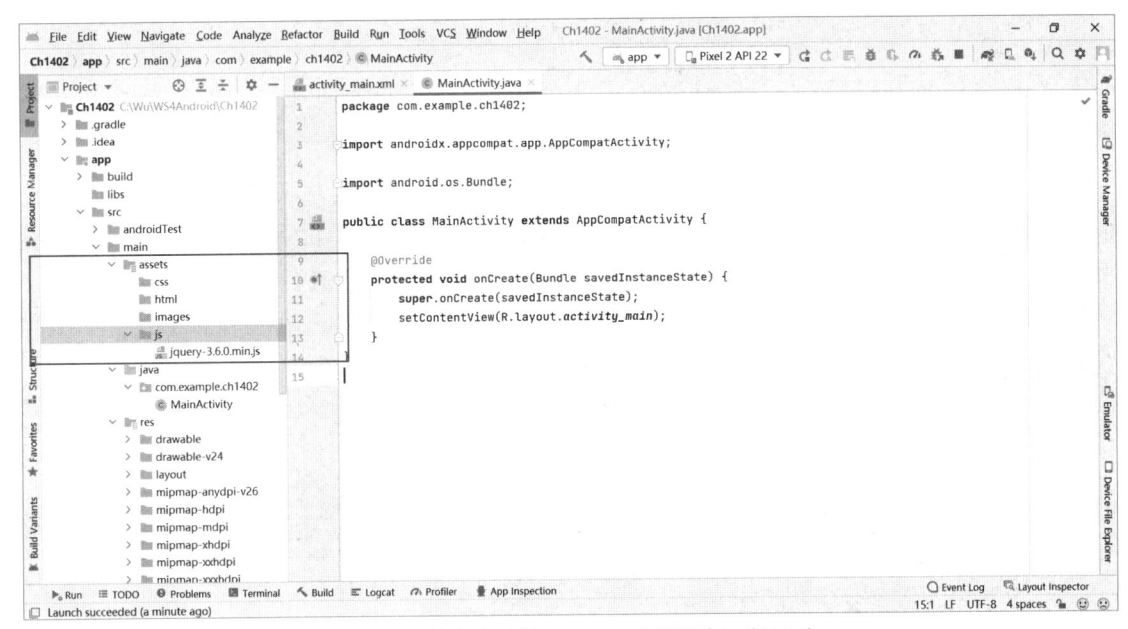

图 14-4　创建完成的 Ch1402 应用程序工程目录

由于需要进行 Android 的原生 Java 代码与 HTML 页面之间的信息交互，因此在 Ch1402 应用程序工程中使用灵活、便利的 jQuery 作为 JavaScript 的工具框架。

修改布局文件 activity_main.xml 中的代码，修改后的代码如下：

```xml
<?xml version="1.0" encoding="utf-8"?>
<LinearLayout
    xmlns:android="http://schemas.android.com/apk/res/android"
    android:layout_width="match_parent"
    android:layout_height="match_parent"
    android:orientation="vertical">

    <EditText
        android:id="@+id/id_edit_text"
        android:layout_width="match_parent"
        android:layout_height="0dp"
        android:layout_weight="8"
        android:hint="input something"
```

```xml
        android:inputType="textMultiLine"
        android:textSize="30sp"
        />

    <Button
        android:id="@+id/id_button"
        android:layout_width="match_parent"
        android:layout_height="0dp"
        android:layout_weight="2"
        android:text="Display in HTML"
        />

</LinearLayout>
```

在 activity_main.xml 文件中，用户可以在 EditText 组件中输入要传递到 HTML 页面中的信息，然后点击 Button 组件，即可启动一个新的 Activity，用于显示 HTML 页面，并且在 HTML 页面中显示输入的信息。修改 MainActivity.java 文件中的代码，修改后的代码如下：

```java
package com.example.ch1402;

import androidx.appcompat.app.AppCompatActivity;

import android.content.Intent;
import android.os.Bundle;
import android.view.View;
import android.widget.Button;
import android.widget.EditText;
import android.widget.Toast;

public class MainActivity extends AppCompatActivity implements View.OnClickListener{
    private EditText edit_text;

    @Override
    protected void onCreate(Bundle savedInstanceState) {
        super.onCreate(savedInstanceState);
        setContentView(R.layout.activity_main);

        edit_text = findViewById(R.id.id_edit_text);
        Button button = findViewById(R.id.id_button);
        button.setOnClickListener(this);
    }
```

```
    @Override
    public void onClick(View view) {
        String info = edit_text.getText().toString();
        if (info.isEmpty()) {
            Toast.makeText(this, "Please input something", Toast.LENGTH_LONG).show();
            return;
        }

        Intent i = new Intent(this, HTMLActivity.class);
        i.putExtra("message", info);
        this.startActivity(i);
    }
}
```

在 MainActivity 类中,首先监听对按钮的点击事件,然后获取在 EditText 组件中输入的信息,最后启动 HTMLActivity。为此,在 Ch1402 应用程序工程中新建一个名为 HTMLActivity 的 Java 类文件,修改其布局文件 activiry_html.xml 中的代码,修改后的代码如下:

```xml
<?xml version="1.0" encoding="utf-8"?>
<LinearLayout
    xmlns:android="http://schemas.android.com/apk/res/android"
    android:layout_width="match_parent"
    android:layout_height="match_parent"
    android:orientation="vertical">

    <WebView
        android:id="@+id/id_web_view"
        android:layout_width="match_parent"
        android:layout_height="match_parent"
        />

</LinearLayout>
```

修改 HTMLActivity.java 文件中的代码,修改后的代码如下:

```java
package com.example.ch1402;

import androidx.appcompat.app.AppCompatActivity;

import android.os.Bundle;
import android.webkit.WebSettings;
import android.webkit.WebView;
import android.webkit.WebViewClient;
```

```java
public class HTMLActivity extends AppCompatActivity {
    private WebView web_view;

    @Override
    protected void onCreate(Bundle savedInstanceState) {
        super.onCreate(savedInstanceState);
        setContentView(R.layout.activity_html);

        web_view = findViewById(R.id.id_web_view);

        //点击超链接，不使用浏览器打开该超链接，直接在 WebView 组件中打开该超链接
        web_view.setWebViewClient(new WebViewClient());

        //使 WebView 组件支持 JavaScript
        WebSettings settings = web_view.getSettings();
        settings.setJavaScriptEnabled(true);
        //允许 HTML 页面访问 assets 目录
        settings.setAllowFileAccess(true);

        //传送数据到 HTML 页面中
        String mess = this.getIntent().getExtras().getString("message");
        transfer_data_to_html(mess);
        //加载 HTML 页面
        web_view.loadUrl("file:///android_asset/html/display.html");
    }

    private void transfer_data_to_html(String message) {
        HTMLSupport html_support = new HTMLSupport(this);
        html_support.setInfo_android_to_html(message);

        //添加 JavaScript 交互接口，并且指定 JavaScript 中 window 对象的属性名称
        web_view.addJavascriptInterface(html_support, "android");
    }

    @Override
    public void onBackPressed() {
        if (web_view.canGoBack()) {
            web_view.goBack();
        } else {
            super.onBackPressed();
```

第 14 章 Android 和 HTML5 的混合开发

```
        }
    }
}
```

在 HTMLActivity 类的 onCreate()方法中，首先获取 WebView 组件的引用，然后设置 WebView 组件的相关属性，代码如下：

```
web_view = findViewById(R.id.id_web_view);

//点击超链接，不使用浏览器打开该超链接，直接在WebView组件中打开该超链接
web_view.setWebViewClient(new WebViewClient());

//使WebView组件支持JavaScript
WebSettings settings = web_view.getSettings();
settings.setJavaScriptEnabled(true);
//允许HTML页面访问assets目录
settings.setAllowFileAccess(true);
```

最后设置要向 HTML 页面传递的信息，并且使用 WebView 组件打开指定的 HTML 页面，代码如下：

```
//传送数据到HTML页面中
String mess = this.getIntent().getExtras().getString("message");
transfer_data_to_html(mess);
//加载HTML页面
web_view.loadUrl("file:///android_asset/html/display.html");
```

在上述代码中，关键代码如下：

```
transfer_data_to_html(mess);
```

transfer_data_to_html()方法是一个自定义方法，主要用于向 HTML 页面传递用户在输入框中输入的信息，该方法中的代码如下：

```
private void transfer_data_to_html(String message) {
    HTMLSupport html_support = new HTMLSupport(this);
    html_support.setInfo_android_to_html(message);

    //添加JavaScript交互接口，并且指定JavaScript中window对象的属性名称
    web_view.addJavascriptInterface(html_support, "android");
}
```

在 transfer_data_to_html()方法中，首先创建一个 HTMPSupport 对象 html_support，然后使用 html_suppor 对象调用 setInfo_android_to_html()方法，传递一条信息到 HTML 页面中，最后通过以下代码将 html_support 对象传递给 WebView 组件，并且指明在 HTML 页面文件的 JavaScript 代码中，可以使用 window 对象的 android 对象属性访问 html_support 对象中的属性和方法。

```
web_view.addJavascriptInterface(html_support, "android");
```

在 Ch1402 应用程序工程中新建一个名为 HTMLSupport 的 Java 类文件。在 HTMLSupport 类中，使用@JavascriptInterface 标注可以被 JavaScript 代码访问的 Android 原生方法，用于在

Android 的原生 Java 代码与 HTML 页面之间进行信息交互。HTMLSupport.java 文件中的代码如下：

```java
package com.example.ch1402;

import android.content.Context;
import android.content.Intent;
import android.webkit.JavascriptInterface;
import android.widget.Toast;

public class HTMLSupport {
    private final Context mContext;
    private String info_android_to_html;
    private String info_html_to_android;

    public HTMLSupport(Context context) {
        mContext = context;
    }

    @JavascriptInterface
    public String getInfo_android_to_html() {
        return info_android_to_html;
    }

    public void setInfo_android_to_html(String info_android_to_html) {
        this.info_android_to_html = info_android_to_html;
    }

    public String getInfo_html_to_android() {
        return info_html_to_android;
    }

    @JavascriptInterface
    public void setInfo_html_to_android(String info_html_to_android) {
        this.info_html_to_android = info_html_to_android;
    }

    @JavascriptInterface
    public void showToast(String str) {
        Toast.makeText(mContext, str, Toast.LENGTH_LONG).show();
    }
}
```

```java
@JavascriptInterface
public void startNewActivity(String message) {
    Intent i = new Intent(mContext, ThirdActivity.class);
    i.putExtra("message", message);
    mContext.startActivity(i);
}
}
```

通过以下代码打开 display.html 文件表示的 HTML 页面。

```
web_view.loadUrl("file:///android_asset/html/display.html");
```

在 display.html 文件的 JavaScript 代码中，可以访问 html_support 对象的方法。display.html 文件中的代码如下：

```html
<!DOCTYPE html>
<html lang="en">
<head>
    <meta charset="UTF-8">
    <link rel="stylesheet" type="text/css" href="../css/style.css">
    <script type="text/javascript" src="../js/jquery-3.6.0.min.js"></script>
    <title>Title</title>
</head>
<body>
    <img src="../images/butterfly.png"/>
    <div id="content"></div>

    <script>
        var message = window.android.getInfo_android_to_html();
        $("#content").after("<div id='android'>" + message + "</div>");
        $("#android").click(function(){
            var str = $(this).text();
            //window.android.showToast(str);
            window.android.startNewActivity(str);
        });
    </script>

</body>
</html>
```

在 display.html 文件的 JavaScript 代码中，使用以下代码获取从 Android 的原生 Java 代码传递的信息。

```
var message = window.android.getInfo_android_to_html();
```

然后将获取的信息通过<div>标签显示在 HTML 页面中。此外，在 HTML 页面中，当用户点击显示的信息时，会调用 http_support 对象的 startNewActivity()方法，从而启动 Android

的一个新的 Activity，并且将用户点击的信息传递给 Android 的原生 Java 代码，代码如下：

```
$("#android").click(function(){
    var str = $(this).text();
    //window.android.showToast(str);
    window.android.startNewActivity(str);
});
```

因此，在 Ch1402 应用程序工程中新建一个名为 ThirdActivity 的 Java 类文件。ThirdActivity 的布局文件 activity_third.xml 中的代码如下：

```xml
<?xml version="1.0" encoding="utf-8"?>
<LinearLayout
    xmlns:android="http://schemas.android.com/apk/res/android"
    android:layout_width="match_parent"
    android:layout_height="match_parent"
    android:orientation="vertical">

    <TextView
        android:id="@+id/id_text_view"
        android:layout_width="match_parent"
        android:layout_height="match_parent"
        android:textSize="30sp"
        android:textAlignment="center"
        />

</LinearLayout>
```

这个布局文件很简单，只包含一个 TextView 组件。

ThirdActivity.java 文件中的代码如下：

```java
package com.example.ch1402;

import androidx.appcompat.app.AppCompatActivity;

import android.os.Bundle;
import android.widget.TextView;

public class ThirdActivity extends AppCompatActivity {

    @Override
    protected void onCreate(Bundle savedInstanceState) {
        super.onCreate(savedInstanceState);
        setContentView(R.layout.activity_third);

        String message = getIntent().getExtras().getString("message");
```

```
        TextView text_view = findViewById(R.id.id_text_view);
        text_view.setText(message);
    }
}
```

在 ThirdActivity 类中，使用以下代码获取从 HTML 页面传递过来的信息。

```
        String message = getIntent().getExtras().getString("message");
```

然后将这条信息显示在界面的 TextView 组件中。此外，为了突出显示 display.html 文件表示的 HTML 页面中的信息，定义了一个 CSS 文件 style.css。sty.css 文件中的代码如下：

```
div {
  text-align: center;
  color: red;
  font-size: 40px;
}
```

运行 Ch1402 应用程序，进入启动界面，如图 14-5 所示。在图 14-5 所示的启动界面的输入框中输入"hello world"，点击 DISPLAY IN HTML 按钮，即可将用户在输入框中输入的"hello world"传递到 HTML 页面中，如图 14-6 所示。在图 14-6 所示的界面中点击文字"hello world"，即可将被点击的文字"hello world"传递到一个新的 Activity 中，如图 14-7 所示。

图 14-5　启动界面　　　图 14-6　将信息从 Android 的 Activity 中传递到 HTML 页面中　　　图 14-7　将信息从 HTML 页面中传递到 Android 的 Activity 中

至此，Ch1402 应用程序实现了将信息从 Android 原生 Java 代码传递到 HTML 页面的功能和将信息从 HTML 页面回传到 Android 原生 Java 代码的功能。

14.4 同步练习

修改 14.3 节中的案例应用程序，修改 display.html 文件中的代码，使其在 HTML 页面中显示从 Android 原生 Java 代码传递过来的信息，并且显示一个输入框和一个按钮，用户在输入框中输入信息后，点击按钮，即可启动一个新的 Activity，并且在新的 Activity 中显示用户在输入框中输入的信息。